# Triumph of the Intelligent

# Books by Seymour W. Itzkoff

## The Evolution of Human Intelligence
### An Argument in Series

# Triumph of the Intelligent

the creation of *Homo sapiens sapiens*

*by*

Seymour W. Itzkoff

Smith College

**Paideia**
Ashfield, Massachusetts

Published in the United States by
**Paideia Publishers**
Box 343, Ashfield, Massachusetts 01330

**Library of Congress Cataloging in Publication Data**
Itzkoff, Seymour W.
  Triumph of the intelligent.
  (The Evolution of human intelligence ; 2)
  Bibliography: p.
  Includes index.
  1. Human evolution. 2. Intellect 3. Genetic psychology 4. Social evolution I. Title II. Series: Itzkoff, Seymour W. Evolution of human intelligence ; 2.
GN281.4.I894        1985        573.2        84-19110
ISBN   0-913993-01-8

*To the Madonna of the Danube*
*Fruition of an ancient biological heritage*
*Bearer of civilization's fragile charm*

# Contents

## PART 3: Arrival

# Diagrams and Plates

(located after page 122)

# Preface

*Triumph of the Intelligent, the creation of Homo sapiens sapiens,* is the second in what will be a four-part series on "the evolution of human intelligence." This book follows closely on its more broadly-focused predecessor, *The Form of Man, the evolutionary origins of human intelligence.* As such, the present book is a development and articulation of my basic argument on the evolutionary processes that resulted in the creation of our species. The style of writing and the narrative approach are here somewhat different, I hope more in tune with the needs of a nontechnical readership. Naturally, I hope that it bridges the requirements of being factually clear as well as communicating the unique and powerful dynamics that produced that peculiar animal, *Homo sapiens sapiens.*

*Triumph of the Intelligent* has been freed of most of the citations that will be found in *The Form of Man.* Instead, I have appended at the back of the book a chapter-by-chapter bibliography, which sets forth recent and general writing that supports my argument at the respective points in the book. The reader may refer to *The Form of Man* for a complete and detailed citation. The complex nature of the evolutionary issues argued in that book necessitated a more comprehensive scientific and philosophical study. In *Triumph,* I wish to communicate the power and mystery of this completely natural process of evolutionary extrusion. Thus I hope that it will both stand on its own merits as a book and still further the message in the evolution of human intelligence that was previously set forth in *The Form of Man.*

As always, for guidance and sustenance, thank you, PSI. My colleagues Stanley Rothman, of the Department of Government at Smith College, and Guenter Lewy, of the Department of Political Science at the University of Massachusetts, have been extremely supportive and helpful. I thank them for their friendship. A word of appreciation to Rita Smith and Deepali Muthana for the care and conscientiousness with which they assisted me with the manuscript. Finally, I acknowledge with gratitude the tangible assistance of the Earhart Foundation, Ann Arbor, Michigan. Its research fellowship grant made it possible for me to complete the work on *Triumph of the Intelligent*.

# PART 1

*Homo:*

the evolutionary model

# I

# About the Unspeakable

Every historical period has its crisis issue. Often it lies below the surface of acknowledgment and discussion. People work around it; they dissemble about its causes and effects. Who can clearly state why it is that at any one moment in history we are incapable of breaking with convention to face that which will burn us, that problem that we don't want to confront? Today that issue is human intelligence. Why are human beings endowed so variably with intelligence and with such diverse results?

Our world is shrinking. We are part of an international scene of common technological and economic demands. Nations once backwaters can leap into the forefront of economic competition. Those once mighty fall; the hopeful and assured collapse in a heap of assumed expectations. Differences between individuals, ethnic groups, nations, are eliminated in the grinding competition of survival in a small, overcrowded world. No longer can a nation hide behind the facade of cultural differences and varying historical points of development. How do we explain events? Can we control a destiny now out of our grasp? What keys are we missing?

At one time, to speak about sexuality or human aggression was taboo. We have overcome our censorships of various discussions—of the heliocentric theory, of the drives of a libido, even about the possible nonexistence of a deity. However, that individual destinies or national aspirations might in some way be determined by personal intelligence, by corporate disciplines and skills cannot yet be faced.

Living as we do in a world of enormous changes produced by brilliant, inventive, aggressive individuals and nations, our shadowy individuality appears to be hurtled into the thin air of events by purely extrinsic or accidental environmental factors. We are victims of circumstance, deprived of autonomy, and thus responsibility. Is this really true? Is there indeed some mysterious force out there that would deny us our individuality, prevent us from shaping the world and destiny as we work with our fellows?

This book is about the unspeakable. It is about the evolution of intelligence in man, how it became the key to man's advance into dominance, how it thence became an obituary notice for rivals, and how in ever wider concentric circles, it increasingly distanced man's rivals from man himself. For understanding the evolution of man, though we could focus on sexuality, or hunting, or toolmaking, in order to see the process properly and in a context applicable to today, we must focus on human intelligence.

The key word is *human*, for we have heard and read too much science fiction of late, proclaiming how like rats or chimpanzees we are. These models are convenient for conjuring up clever fairy tales about us, but they square neither with the reality that we see around us today vis a vis human behavior nor with the scientific facts and probabilities of the evolutionary process as it effects our line of progressive anthropoids.

This is the purpose of the *Triumph of the Intelligent*. I intend to show that man has inherited and perfected one of the most basic of the adaptations that animal life has exhibited in its struggle against energetic degradation since the time life itself appeared on this earth. The path taken by the adaptation for intelligence has been slow and circuitous. Intelligence is a complex adaptation necessitating not only much neurological baggage but also the right creature at the right ecological moment. With the ebb and flow of animal life, it has bided its time, riding piggy-back, so to speak, on those creatures that could best use its quixotic self-regulating character.

As I will point out, intelligence was made a necessity for those creatures that lived at the very center of the selective contest. These creatures were under constant environmental pressures as the world changed. They mutated often, were frequently able to take advantage of change, and most interesting, they stayed relatively unsuccessful, thus not getting themselves too deeply into an adaptive rut, as did the snails, starfish, clams, or worms.

In arguing for the dependence of *Homo*'s adaptive success on intelligence, I will make the following claims:

1. Man is not directly descended from a recognizable ancestor of any contemporary ape or monkey types. More controversially, the morphological evidence seems to point to an originating time in the Oligocene, perhaps thirty million years ago, since man has both ape- and monkey-like characteristics. That he was a singularly unsuccessful primate seems probable when we consider those basic adaptations that he carried with him for so many millions of years into the present: intense familial bonds, language, delicacy of structural (bone) features, large brain, partial bipedalism.

What is interesting about man's long, quiet, precarious journey of inconspicuousness until he moved into the late Miocene savannahs some ten to eight million years ago is that this pattern is in the classic tradition of such evolutionary leaps. The ancient therapsid mammals existed for almont seventy-five million years—under the noses and dominance of the dinosaurs—as they perfected their unique adaptations and were transformed from a line of reptile losers into the veritable mammalian destroyers of their dinosaur competition.

The evolutionary road leading to *Homo sapiens sapiens* has been long, plagued with much travail and insecurity, only relatively sudden victory. Viewed in this way, the important evolutionary differences between the hominids and their pongid cousins can be understood and appreciated.

2. *Homo* did not emerge from the forests all alone. He was part of a general movement of preadapted anthropoids that also used this new ecological opportunity. So late in the history of the mammals did this movement take place (late Miocene) that the tongue-and-groove adaptive dependencies of lion and wildebeeste, for example, were well fixed and did not engage the selection of the hominids.

Man had to struggle, learn how to make a living, and move long distances in search of food. In the process he grew in size; so did his already ample brain. As he multiplied with modest success over the millions of years, he came into competition with his own kind. Many ancient forms of these progressive open-country apes existed at the same time as the australopithecine man-apes, at first quite successful, well-adapted intermediate creatures.

From a tiny, defensive forest creature, *Homo* gradually evolved into a larger, more aggressive, highly intelligent, semi-carnivore. The disappearance of large numbers of ape species during the Pliocene (seven to two million years ago) and the survival of only the more specialized jungle apes of today point to the pressure exerted by both *Homo* and the australopithecines. Next, the australopithecines, highly able, long-existing creatures, were subjected to *Homo*'s aggressive expansion. First, the smaller open-country omnivore *Australopithecus africanus* disappeared (ca. one million years ago). Finally the jungle herbivore *Australopithecus boisei* left the evolutionary scene (ca. 500,000 years ago).

The war between the hominids was not yet over.

Wandering over the eastern hemisphere, these relatively instinctless creatures doubtless had no qualms about inflicting their highly charged energies upon their own kind. Cultural separations allowed for a sense of "they" and "we." Where genocide was incomplete, miscegenation took place, allowing for parallel and relatively uniform rates of evolution, securing the genetic relationship between humans until the very end.

3. Man's intelligence was different. Heretofore intelligence had been honed to a task: the wiliness of the bluejay, the stealth of the panther, even the playfulness of the chimpanzee had clearcut adaptive uses. Keeping in the background—which allowed for only an unspecialized and evasive intelligence—man tore loose from the close honing selective process. Man had no special attack organs—no fangs, claws, or hooves. He was relatively slow afoot, did not camouflage easily. There is evidence for his relative independence of the food chain, his immunity to attack by predators aside from the occasional encounter with tiger or lion.

Human intelligence expanded because of the raw need for survival in a lonely, instinctless existence. Even in modern times, humans have had to struggle to make a living in the tribal world of feast and famine. Imagine what this struggle must have been a million years back. Even then, however, intelligent individuals might look forward in time to predict what could happen on the basis of past events. In picking up clues from the natural world that might help them use their time and energy more economically for hunting or gathering, they thereby helped the family and the band.

Where intelligence probably became truly selective was in the struggle between rival human groups and the consequent intratribal process of social selection—who got which mates, whose children were born or not. This process of selection could not have taken place without the appearance of mutations for higher intelligence. They did come, simply because higher intelligence had worked selectively before and the tradition of mutations for larger brain size had been reinforced by natural selection. George Gaylord Simpson called this well-established biological principle "orthoselection." Through orthoselection, genes for higher intelligence of a very general sort came forth, were absorbed into human societies, and tilted human evolution in its inevitable direction.

4. Culture is an ancient characteristic of our line. Even the australopithecine man-apes made stone tools. Ralph Holloway has discovered that the more pongid-like *Australopithecus boisei* seems to have had a Broca's area, a language center on the left side of his brain. If this creature made tools, could speak a few words utilizing a rudimentary syntax, there is reason to believe that other elements of culture could have also existed: dance, social structure, technology.

That such a lifestyle would be possible for the australopithecines argues that the genus *Homo* must also have had a nascent cultural tradition, perhaps as far back as five or six million years ago. The earliest known human, *Homo habilis*, of about two million years ago, had a brain quite a bit larger than his australopithecine first cousin. Of course it was a quintessentially human brain.

As we trace the record forward in time, we see a parallel between the quality of toolmaking and brain expansion. The reasonable assumption must be that culture is a basic characteristic of the hominids. However, the intellectual qualities of the culture vary directly with the size of the brain and the corporate intelligence of those human groups that share a genetic and social tradition.

Don't be surprised that all human groups have culture. Do be surprised if all these groups have equally developed, sapientized cultural achievements.

5. We are an extremely variable yet interbreeding species.

When *Homo erectus* wandered out of Africa over a million years ago, he took with him a tradition of intellectual and cultural variability plus the ancient large mammal characteristic of migratory interfertility. This latter meant that even though humans would be separated by thousands of miles and thousands of years, when the genes would meet again, they would prove to be interfertile, even though the external physical features—e.g., those of dogs, cattle— would be quite different.

The variability of the brain both in structure and size arose presumably because this organ had been subject to positive selection for so long a time. Because what natural selection existed was carved out of internal social dynamics, the patterns of selection in each human band could be somewhat different. The fact that so many biochemical combinations control the growth and patterns of brain structure indicates that slight variations can exist between extremely close blood relations and large variations between relatively isolated groups of humans. The brain simply evolved along a wide diversity of lines. The result is obvious in culture and personality, in the quality and the quantity of intelligence.

What has existed for so long within the human genus— variability—should hardly be expected to disappear sudden- ly. One brief glance at late-twentieth-century humanity, what with intense stereotyping forces in communications and transportation, technology, and political centralization, should argue for this truth.

# II

# Intelligence in Search of an Animal

### Fighting Energy Degradation

The billion-years-old seas were suffused with hydrocarbons—large stringy molecules composed of hydrogen, nitrogen, oxygen, and carbon atoms. They hissed into being with the explosion of lightning bolts. Then gradually they melted away into the warm, thick oceans. In other places, erupting magmas from the incendiary interior of planet Earth poured forth, extruding geysers of steam and causing momentary bubbling of the surrounding waters.

The molecules again swirled, now into the nearby seas, thin membranes separating the highly-charged organic materials—like globules of fat in a bowl of chicken soup—from the more prosaic aqueous surrounding. Soon they too cooled, though occasionally putting off death and dissolution by merging with adjoining hydrocarbons, maintaining a momentary warmth against the inevitable energetic degradation. The second law of thermodynamics and the principle of entropy, however, were inexorable. Energy, according to the second law, would be distributed both evenly and uniformly throughout the surrounding environment. Opposing this trend, life made its first hesitant "illegal" beginnings on our planet.

Above absolute zero (-273° Centigrade), all environments or objects contain within themselves a measure of energetic activity. Place two containers of water next to each other in an enclosed space, one with a temperature of 100° Fahrenheit, the other of 200°; it is obvious which will get warmer. No, the 200° F. water will not take heat from the 100° water.

Life, however, runs counter to this universal thermodynamic principle. It does not violate it, because the principle does not deny momentary, if billion-year, eddies in the irresistible flow of energy downward toward uniform cold. In the end, all of life too will succumb, perhaps to begin somewhere else, maybe on the edge of the Milky Way, maybe far beyond.

Somehow the primeval infusion of energy into the seas was of longer duration, not as destructive or violent as the lava nor as short or brittle as the lightning. The molecules endured, joined, divided, grew larger, became more efficient internally in building up the membrane, absorbing from the rich soupy seas more organic, energy-rich foods, and persisted. Perhaps life came out of the blazing green-gold of the sun, and from that permanent, nurturing source of energy eventually gave birth to a billion species less immediately dependent on the sun, more ravishing and cannibalistic toward their neighbors.

Here, predatory intelligence began, in that race against cosmic time's appointed moment of dissolution and nothingness. Life ran counter, dynamic and desperate, to beat the cold outside by metabolizing more energy than it gave off. It has never changed. The point of life is to burn fuel, to excrete the waste products (our inefficiency in the use of this fuel), to reproduce our kind and bring our progeny to repoductive maturity, then to fade into individual nothingness, passing on the baton of time and history.

**Motility**

From that enigmatic but critical moment when living

things were catapulted into a systematic race to acquire more energy-laden nourishment from the surroundings, the various ingenious strategies of survival that are chiseled into the evolutionary record came forth. Some billion years later, algae, mosses, lichens, tiny one-celled droplets still exist whose solutions to the problem of survival have remained and will probably remain true to the test of natural selection until close to the end of our story here on earth. The combinations were economical and conservative, not overly demanding or competitive in the thrust for maintenance or renewal. Somehow other solutions soon took form.

Our knowledge of natural history today is such that we don't have to spell it out in detail. The environment in which life began—the seas—determined the conditions of development and evolution. The great biochemical breakthrough of photosynthesis whereby living things absorb the energy of the sun and transform it into free oxygen made so much possible. In return for this security and permanence of energy, the plants have remained the great conservationists. As they sank their roots into the watery muck or, later, the upland granites, they represented the other energy decision—away from learning or intelligence.

From other patterns of adaptation—giganticism, dwarfing, fecundity, conservation—ecological specialization can be observed and chronicled over the eons with each new structural development. The animals, perhaps a more primeval solution to the challenge of entropy, took off in their cannibalistic search for organic food in a wider arc. Here intelligence began, in movement—in *motility*. Whereas the plants sent their tendrils upward, in search of inorganic sustenance, the animal parabola of adventure was more horizontal.

Even given such basically different solutions, animal adaptations have assumed parallel patterns in their varying stages of development over the hundreds of millions of years.

The water scorpion or eurypterid *Pterygotus* of four hundred million years ago was almost as large as a Cadillac. It would preside at the head of the living equation for a few hundred millennia. Like all comet-like superstars, it soon disappeared, now only a trace of mineralized substance at the bottom of some ancient dried-up sea bed.

It has been a long hegira into time. Many have taken those adventurous steps into the unknown. Most have fallen, disappeared without leaving even a fossilized trace. All experiment has its successes and failures. At each point in this game of molecular roulette, whenever a new pattern of life and digestion presented itself, most animal forms have heaved a sigh of relief for sanctuary. They have settled in, preferring security over movement and adventure. We see these types even among our human friends and neighbors.

The barnacles, sponges, and corals were once full of moxie and possibility, but, rather than advancing, they just barely hung on, even "repressing" experiment. Other coelenterate adventurers evolved—into starfish and sea urchins (now they are echinoderms)—and became a bit more mobile and adaptable. The clams and snails, also typically sedentary, derived from ancient, rarely changing forms, the latter from the ammonites, once among the most glorious and numerous dwellers of the sea bottoms. They hit on something good—quiet, unvarying conditions of life. Not asking too much, they needed only the security and the protection of a rarely disturbed environment. Five hundred million years later, they are still here.

With humans around now, polluting the waters, these ancient dwellers have been put on suffocating notice. Trawlers suck the sea bottoms scooping up the clams, who wind up being twirled at the end of a small fork in horseradish sauce. They hadn't planned on that exigency.

## Instability

It is simplistic to claim from the foregoing that the sure

path of the adaptation of intelligence is instability. It is true though that the many evolutionary "dead ends," the quiet, safe niches, testify to the "thus far, no farther" paths taken by many forms of animal life. Stability, instability, it is all part of the cosmic game of chance. The wheel has circled for a billion years. Here is our reassurance of regularity, that the conservatives will eventually win out, even after the travelers have been beached, the human animals incinerated in manic follies.

However, with each passing circle of the sun, new niches, new moments have been discovered, and living things have been forced by events to reassess and react to the new. Since the basic physical laws seem to favor the most dynamic energy gatherers, natural selection in its random reflection of these winds of change has also allowed for the chance takers. These are the movers, who are never fully settled in; they have held on, taking their chances with circumstance.

The big successful guys fall hard. Life has had a succession of them. Even the vaunted redwoods, of an incredibly ancient floral line, have come a cropper to the ultimate enemy, man. Over the eons it has paid *not* to be too big a winner; better just hang on, for with each millennium comes another round of opportunity. Nature may offer the big break.

In order to be successful over the long term, creatures have to be somewhat unsuccessful over the short term. One of the requirements is for the line of motile creatures to be ready to move into new ecologies or geographies. To do so creatures may have to change or evolve. This is done through mutations. Simply stated, a mutation of the genes is a product of nature's long history of change and its biochemical wager that change will take place once more. Thus mutations occur without conscious awareness in the carrier, the creature.

Since it has worked in the past that a spontaneously altered biochemical chain has allowed the carrier of the altered chemistry to survive when otherwise it ought not, each of the unstable survivors of evolutionary history carries within it not only a history of mutations but indeed the reality of the creature in process of changing. If those mutations popping forth here and there occasionally help the line to survive, it is because the roulette wheel has coincided with the winds of change. The earth itself is an evolving entity. New continents, seas, and climates constantly present themselves to alter its face and the denizens roaming upon it.

We are not as aware as we probably should be that the hominid line has been one of the most variable and rapidly evolving families among all the animals. It is a product both of our heritage of instability and the fact that it has been useful for our ancestors' survival to be a product of that genetic instability, with which we are still blessed. What happens sometimes, as Harvard University's Ernst Mayr has noted, is that any long successful line finding a secure ecology will over a great period of time be slowed in its rate of mutational change.

The unstable characters will recklessly crawl out on a limb that may get chopped off, while the less ambitious, but now successful, will stay behind and successfully bring their progeny to reproductive maturity. A series of enzymatic buffers will block the frequency of mutational changes or, to put it another way, those members of the line who now have this variation for less ambition will succeed more often in rearing the next generation. This trend toward fewer mutations will be reinforced.

The unstable ones venturing onto the highways of change will still often succeed. If the conservatives are to get hurt, they will eventually fall victim to a genocidal cataclysm, unable to muster the variability that would extricate them from the challenge. The history of life is rarely a chronicle of the sure thing.

## Variability

It is dangerous to be homocentric about evolution, to think of it exclusively in terms of its human outcome. It is clearly tempting to do so. For example, we could speak about rich and self-satisfied societies that have so often reached a peak only to break up and disappear. The stagnant Egypts can be given as examples of the slow rot of society. So, too, the issue of centralization, as in the Soviet Union, comes to mind, a clear case of hardening of the arteries. Control from one center often subjects the entire civilization at all points in its widely separated geography to possible suffocation.

Thus a line of animals, even the unstable, venturesome ones, could be caught in a momentary bind—a localized blight or a sudden even if mild alteration of climates or conditions. If all the members of the line are uniform, they could all uniformly perish. If however, slight deviations of color, structure, size, or behavior occur within the group, perhaps the blow would not be so devastating; some might survive and reproduce.

Thus the principle of variation might be reinforced, but so would the principle of interfertility, in which slight variants could still reproduce with the main or even with other borderline types. Being on the edge would thus not mean permanent exile from the main family branch. Often the variants do go off, far enough in space or long in time to become separate species and thus by definition no longer interfertile with the main group.

So too, a healthy society allows a wide variety of communities, patterns of life and beliefs to coexist. One never knows which one may be needed in a time of travail, sometimes for the survival of the society as a whole. Over the long run, like good Boy Scouts, successful forms of life have had built into them the genetic maxim: "Be prepared."

# Of Time and Learning

The psychologists define learning as "a change in behavior as a consequence of experience." Thus, in theory, plants do not learn. What "behaviors" they may seem to have, we call tropisms—attractions and repulsions of unvarying response and rhythm. By contrast, were we to place a light near the lowly amoeba or paramecium, it would respond positively by being drawn toward the light. Shut off the light. A few seconds later, turn it on. Did the amoeba swim toward the light somewhat more rapidly? If so, it has "learned."

Naturally, most learning is predictable and determined by genetic programming—this we call instinctual reaction—for Mother Nature doesn't yet trust the creature to make decisions by itself. Learning is made possible and necessary by the animal's motility. The more sedentary a creature is, the more it approaches the plant-like state. Some sedentary animals have surprises for those who approach unwarily. However, in general, the less a creature moves around or charges into foreign or changing environs, the less it needs to react variably to different external experiences.

Creatures on the prowl, if unstable or harried to the point of extreme defensiveness, need to learn to react to different stimuli. It is for these creatures, usually on the main line of mutational change both for behavior and structure, that learning becomes crucial. For those that are unstable, Mother Nature would do better to provide a repertoire of versatile behaviors with diverse sensory inputs rather than one of slow structural mutational changes. The more variability in a line of animals, presumably the more diverse the structural responses.

The catch is that such responses will take generations to produce in enough quantity ever to overcome the negative external conditions. On the other hand, a repertoire of variable behaviors spread across a diverse population could be useful for survival purposes in a matter of weeks or months. Thus those creatures in the main line who ought to be constant mutators might, even better, be learners.

This variability of response explains why the young of such advanced creatures as lions, dogs, or even cattle are far friskier and more playful than their parents. They are, while they are still young, in a sense trying out their repertoire of behaviors to test the environment, to see what is reinforced, what necessary for the orders of the day.

Natural selection has thus given to those who have chosen to evolve dangerously a complex set of neurological information gatherers/integrators. It is much easier to create giant Baluchitherms or mole-like midgets or creatures that sow their seed wildly and profusely, like salmon or rabbits. The evolutionary route for learning takes much time, for here the genes must be taught biochemically to surrender their behind-the-scenes control to stage front, where the action is.

For learning to develop as a really useful adaptive strategy in an animal, time and the proper physical structures are necessary. Sensory organs will have to transmit information more complex about the outside world than gross physical impacts or temperatures. These information bits will then have to be organized to provoke a decision-making process—react or don't react—about what is important or irrelevant for survival. A control system has to exist that is almost as infinite as that of an electric stove. At what point does a shadow, a noise, or a smell evoke the snarl, flight, or freeze? This is now complicated stuff. After having experienced several hundred million years of surviving on the brink, the genes have enoughed stored information and experience—reflected in the basic makeup of the creature for the neurological, behavioral, and thus learning option—to be adaptive for an animal that has been long tested and is still around.

It is odd, isn't it, to think that intelligence came into its own as an adaptation for escape, defensiveness, even bare survival? Great size, strength, ferocity, or speed are specializations that had to detract from the enhancements for learning and reacting. Yet, when we think of man, the intelligent animal par excellence, we cannot but wonder, is this a defensive creature? How has evolutionary intelligence gone astray? Why would nature now allow the use of intelligence for the inflicting of such havoc on life and the physical environment?

## Intelligence and the Land

What is youth, what, age? The universe is perhaps twenty billion years old, our own planet, perhaps four. Life began between one and two billion years ago. It can be said that intelligence could evolve only while the earth was still young. It needed time, slow and steady time. While the earth was yet pulling out new tricks to throw at the scrambling forms of life, change and stability remained in a healthy tandem of tension and balance. While flexibility remained, the lessons of biological intelligence could yet be mastered.

Life's many adaptations have been and are seen in homologous creatures. The flying reptile pterodactyl, the bird, the bat, all soar through the heavens. In the water, all the great forms, invertebrates, fish, reptiles have flourished, even our own returning mammals, until man took off after the whales. Giants, midgets, camouflaged obscurities have had their moments. All the while, the intelligent have steadily pushed forward.

They speak of the shark, a primitive fish with a large brain. There is also the great-beaked squid. Many years ago, Carl Owen Dunbar noted that this latter ancient invertebrate was an inordinately intelligent animal that would have been a formidable match for many of his more modern aqueous competitors. In this example we see how important for the expression of intelligence were structure and environment. The water gave animals little opportunity to hide away to develop those new morphologies that would have enabled them to make a competitive leap over their rivals.

There are undoubtedly many other potentially intelligent, if primitive, aqueous creatures, including the run of-the-mill gill-breathing teleosts, salmon, trout, or cod. As the competition heats up in every ecology, it is natural that forms of intelligence that are still malleable and defensive can be pressed toward size and ferocity too. However, at that point, the book of intellectual progress closes.

In the Devonian period, about 325 million years ago, long after the squid and the sharks had marked off their territory, an opportunity arose. Great geological changes in the lands and seas had caused enormous stress in the status quo. Plants now softened the landscape of upland rocks, providing an inviting scene. Here, in a setting of drying lakes, an obscure, retrogressive, bottom-dwelling fish, still harboring its ancient air sac, reminiscent of that old relationship of life and the interface of air and sea, made its involuntary leap.

Lobed-finned, for stability at the water's bottom, it flip-flopped from mud hole to mud hole gasping air, gradually making its commitment not to return to the deep. It would take millions upon millions of years before the umbilical connection with the seas would be severed. The amniotic egg would have to come into being. With homeothermy (internal regulation of heat) and viviparity (internal gestation of the young), the conditions would be established for the mastery of the new environment.

Head-high creatures tend to be sensitized to distant vistas, sounds, and then even smells. The small brain derived from ancient fish, amphibians, and early reptiles, would expand within the skull. It would become a neurological midpoint between the various sensory sources of information. Heretofore slow, the neurological tracings could now be amplified and organized. Highly cerebralized birds would vault from the trees, an extraordinary visual brain creating an entirely new aerial domain for life. On the ground, tiny, conservative reptiles, skittering in and out of the shadows of the giant dinosaurian host would mark time, as neuron and synapse encoded messages of evasion now in both day and night. The alpha waves would rarely go dormant.

Much time would pass before the adaptation of intelligence would make its mark in this new ecology. The birds, brainier and more specialized, limited by their fragile perch, would go their own way. Even with the mysterious demise of the dinosaur, the road would not be easily cleared. Great opportunities on land would still produce giganticism, predation, and stupid, armadillo-like defensiveness. The bear-like cetaceans would flee the competition for the seas, the mouse-like bats would opt for the heights.

In the shadowy, threatening jungles was the furtive one. Like its mammalian therapsid ancestor, tiny and excapist, it was nurturing its genetic possiblities, shuffling the adaptive cards of variability, generation after generation, just to keep going. Here the heritage of intelligent behavior was being quietly reinforced and readied. As the land filled and competitors went at each other, survival on the margin had its value. Don't make your move too soon. Bide your time. The earth was maturing, life reaching a crescendo of change. The brain's moment would soon arrive.

# III

# Odd Anthropoid

## Monkey or Ape

Great moments in time do not always begin with cataclysms. More likely they start with quiet agonies or small sequences of fear, danger, and decision. The origin of the hominids must have been so. We must go back much before the chimpanzees, probably to a time when there were no great apes as we know them today. We will be looking for man in the making, a primate, but unique and true to his special image.

It is critical that we push aside images of human behavior carved from the imaginations of psychologists interpreting rat behavior, trainers of pigeons, or awed observers of the social insects. We often take our special nature for granted, assuming it all developed in a god-given touch of mortality and immortality. It was not, however. It was hard won.

Bernard Campbell, author of *Human Evolution: an Introduction to Man's Adaptations*, seems to think that it came sometime in the Oligocene period, about thirty million years ago. This was when the first true apes were beginning to be identifiable, presumably when the root monkey stock was also branching off. Certain characteristics of our bodies are ape-like, some lean more to the Old World root monkey side; still other characteristics at this point can't be attributed to either, but are perhaps similar to the ramapithecids, whose fossils increasingly reveal a separate evolutionary pathway from most of today's apes and monkeys.

Somewhere in the forests a number of descendants of the prosimians had come sliding down their tree trunks to stay. These tarsiers and lemurs and their ancient ilk had perhaps gained enough bulk and brains to chance it on the ground as well as in the trees. They must have done well, for by the Oligocene these descendants were dominating, pushing their wide-eyed ancestors into the nocturnal dark of extreme specialization and gradual extinction.

Our hands, our use of our thumbs as well as our feet are somewhat monkey-like. Monkeys are less prehensile (able to grab things) with their feet than with their hands. Apes are extremely versatile in this respect. The human male's penis shares more features with the monkey than with the ape. Even the monkey face is far less prognathous. Apes have great jutting chins or beetle brows; the delicacy of the monkey face sometimes reminds one of homunculi.

Both ape and monkey can walk upright, though not with grace or ease. It's easy to see when you go to the zoo that the knuckle-walking of the apes is a specialization that must carry with it an ancient heritage. Contrast the human infant. It will grab a bar or a limb that is off the ground and hold on with its tiny, powerful arms. Then at seven or eight months it is crawling around like a house afire. Finally it stands erect. As one watches the growth and maturity of one's own infant, one sees little glimmer of ape in the process.

Yet our biochemistry is strikingly alike. Vincent Sarich and A. C. Wilson tell us that the existing minor variations in amino acid structure in our cells argue for a human-ape separation of only four to five million years. This is phenomenal. Elwyn Simons and David Pilbeam have tried to push this back to fifteen million years when adjusting for the lengthening of the anthropoid generation. One wonders what kind of biochemical game nature played here in what ought to be a regular time clock of change. For example, David Pilbeam notes that according to the biochemical evidence the laboratory mouse genus *Mus* and the laboratory rat genus *Rattus* should have diverged about thirty-five million years ago, yet fossils of these creatures discovered in Pakistan suggest that they separated only about ten million years ago.

Perhaps the apes grew out of a protohuman stock surprisingly recently. More likely we don't yet have the full picture and meaning of biochemical variation. Too many other aspects of man's unique climb contradict the late divergence hypothesis to pin the timetable of ascent on only one factor—biochemistry—and to label it as the keystone of the mystery.

## Important Differences

The weight of probability points to that ancient Oligocene origin. We must return to the facts of our uniqueness. If we have the largest brain-to-body size of all mammals—and far beyond our anthropoid cousins—then that must have been a factor. We have a peculiar brain too. No other anthropoid has any hemispheric specialization into left- and right-sided separation in function. It seems too basic a structural characteristic to have developed overnight. Also, the human pharynx is, compared to that of the apes, a single structure, with a tongue free at the back of the throat, which allows humans to create a vast variety of sounds. The human baby's tongue is frozen for the first six to eight weeks. At some point, this alteration toward the mature human pattern out of the universal simian structure must have allowed new possibilities of communication.

Whereas chimpanzees and monkeys scream at an extremely high decibel level, 3000 cycles per second, the human voice is usually heard at about our 440 A (the pitch to which a symphony orchestra tunes). A critical evolutionary lesson lies in this parallel development of a brain with a language center rooted usually in the left hemisphere and a pharynx freed from the specialized simian cry. Humans speak in such variable and still conversational tones. How did the change come about, what did it mean for the adaptation and survival of our primeval man?

My guess is that it meant very little in the beginning. The reason I say that is because humans are unspecialized anthropoids except for one morphological feature, the brain. The brain is a huge protuberance in most humans, some say even a hypertrophy. However, though it may have been somewhat larger in the first hominds, it couldn't have been selectively important, because we haven't yet discovered any of our direct ancestors until about twenty-seven million years later, which is quite a bit of selective evolutionary water under the bridge.

Remember, the mammals even in the earliest phases of their evolution 150 million years ago, under bleary dinosaurian eyes, had a larger brain-to-body ratio than the dinosaurs. Unfortunately the mammals were tiny mouse-like creatures and the dinosaurs huge behemoths. The big/little brains couldn't compete with that large simpleton's visual machine.

Another example of the early accidental character of so many of these features, as reported by Stephen Gould, is an Asiatic Oligocene lemur with a pelvis flat enough to be in the hominid position. The human pelvis, although not the most efficient structure for encouraging a bipedal gait, has been shifted around from the slanted, angled simian model. That there was once a lemur with a human-like pelvis, probably existing too late in time to be an ancestral form, at least reveals to us the wide variation in morphological structure that probably was characteristic of a numerous class of creatures—the primates.

On the other hand a flat pelvis hints at an erect stance.

Our tendency toward bipedalism had to come from somewhere. Perhaps it grew out of the up-and-down treetrunk-hopping of the prosimian lemurs and tarsiers of the Eocene period (forty million years ago). Perhaps we came from a more conservative, unspecialized line of evolving prosimians during that even more ancient remove. The word "conservative" is crucial, for whatever derivation our line had, it obviously did not go anywhere for a long time.

It is clear that we cannot look to the monkey for our origin. The monkeys, too, interestingly, originated early, but did not become numerous for a long time because of the ape avalanche then beginning. The monkey advance took on impetus after some monkeys, because of their wanderings, were trapped on the new continents while the Old and the New Worlds were still connected. They then multiplied into what are now the New World monkeys, with their wonderfully prehensile tricks of tail.

In Africa, Europe, and Asia, the Old World monkeys exploded in number and variety after the sudden eviction of the greatly successful apes. This seems to have happened in the Pliocene some seven to five million years ago. What happened there? It is another interesting puzzle, one that we shall address later on.

Remember my claim. Man is not merely one of the apes. He is that, but more. Man is the preeminent mammal, even the prototypical mammal. To understand how he got to his place of eminence and power we don't have to invoke theology or even the random accidents of pongid evolutionary history or climatic and geographic change.

One could never have known what was going to occur when, in the Oligocene, it all started. The forces, the events, the accidents were not atypical of those that had favored other great biological revolutions. Then, it would have been impossible to predict that the outcome would be *Homo sapiens*. (It is still hard to believe as we look out about us and comtemplate the biological wreckage effected by *Homo*.)

Clearly, natural forces did produce our kind. Yes, from such simple and unobtrusive events can come even this preeminent creature.

What I want to emphasize is that the basic evolutionary character of causes and products is not unusual. It had happened before and will happen again. Then, the central character will not be man. Your guess is as good as mine: baboon, mosquito, starling, fungus. It is lurking out there, biding its time, probably on the edge of survival. It may not have to wait too long.

## Sometimes Losers Aren't

The almost failures: A group of fish-like creatures called crossopterygians lived in the seas for many millions of years before the modern teleosts developed. Teleosts are the common mid-depth fish that decorate our dinner plates on Fridays. They are fast, agile, well-adapted creatures. The crossopterygians were lobed-finned fish that lived at the bottom of the inland seas. They hobbled around the bottoms, their fins helping them maintain stability. On occasion they rose to the top to gulp air. About 325 million years ago, they had just about had it. They were played out.

Coming from an ancient line, they probably derived from an early vertebrate creature that existed on the border of land and water sometime after the explosion of oxygen into the atmosphere. They had developed an air sac with which to breathe. With the coming of the great seas, they had been forced to abandon their position of dominance. New fishes had taken over that were highly competent in the deep waters and that had fully developed air bladders, which were used to separate the oxygen from the water.

Poor crossopterygian. It needed air yet was consigned to walking the depths. At least it was out of the traffic in the tremendous surge of the teleosts. Thus it had a chance to endure and hope for some evolutionary breaks. Indeed its luck, and thus ours, was with it. A cousin of the crossopterygian has lasted on the ocean bottom into our own time—the coelacanth, found at various times in the nets of Indian Ocean fishermen. It is a huge fish, as much as five feet in length.

Sometime during the Permian period three hundred and fifty million years ago, the lands began to rise. Inland seas gradually dried up. In this process, great evolutionary cataclysms occurred. Not only fishes but many varieties of water creatures were disturbed. For most, including many teleosts, that flopped from mud hole to mud hole and then to extinction, the news was bad.

For one or more crossopterygian types, it became a heaven-sent opportunity, because, as they flipped and flopped and gasped, they breathed and lived. Like the mudskipper of our own day, that can live both in and out of water, but that is essentially a water creature, this land and water fish now could live a dual life. (Today, however, life is tough for the mudskipper.) There is now plenty of competition in both realms, but three hundred million years ago the lands were empty and the new amphibians simply found that they couldn't go home again.

## The Great Mammalian Evasion

Many millions of years later, a similar and epochal situation occurred. By this time, the Jurassic, two hundred million years ago, the reptiles had evolved into fully adapted land creatures. The amniotic egg (hard shell enclosing liquid content and embryo) was the keystone adaptation and the reptiles and their advanced and sophisticated leading representatives, the dinosaurs, were recklessly loosing their eggs all around the world.

As in all evolutionary processes, there were winners and losers. The loser in this case was a line of ancient reptiles called the synapsids, low-slung and small, that had persisted in eking out a living at the edge of the great advance. Scientists are not yet sure what the circumstances of these reptiles' retardation were. Were they stupider than the dinosaurs, were their reproductive habits less efficient, their egg-laying patterns not completely successful? Were they perhaps merely shrimps in an era that gave the mutation for giganticism a selective premium? At this point, we can only conjecture.

This line hung on, in fact was quite numerous in the variety of types eventually produced. The therapsid reptile successors, however, seem to have been the more able, and they dominated this particular ecology in the era of the great reptiles. Small, from the size of large mice to the size of rabbits or pussycats, they were on the scene for many millions of years, persisting, but not prospering. Subject to pressures from without, they manifested a steady rate of mutations for internal change. The dinosaur engaged the obverse of the coin, mutating with great relish as almost any change in structure seemed to be an adaptive plus. Thus our children can enjoy studying the vast variety of dinosaur types that make up that hundred-million-year spree of squooshy plenitude.

As long ago as 150 million years ago, a different line of therapsids began to develop. Evidencing new bone structures in the auditory areas, a brain that was retreating into the skull cavity equidistant from the three distance receptors (eye, ear, and nose), showing signs of internal viviparity and external milk-secreting glands, these creatures were apparently under heavy selective pressure. Clearly, their way of life was pushing them selectively into new morphological and behavioral ecologies.

That this worked is seen in the fossil evidence. The new mammals endured. The other therapsids did not. Long before the last dinosaur had breathed his weary *addio*, the therapsid reptiles had disappeared. Why? Could dinosaurs have done them in? Certainly the presence of the dinosaurs blocked egress from their yet ample ecological redoubt. I am skeptical that the big fellows did them in. I would bet on the mammals. In nature the keenest competition is between neighbors, even cousins. I dare not say brothers. Creatures inhabiting similar ecologies usually battle for elbow room. Unlike what occurs in the chain of life—where because of their dependence on the smaller or weaker animals, the bigger animals do *not* kill off their food supply—the battle between cousins is often deadly to extinction. My suspicion is that the mammals had developed enough useful defensive adaptations to live out their nocturnal vigil in caves or trees to utilize their higher metabolic rate to advantage.

Homeothermy (warm-bloodedness) allowed a creature to function in a variety of climatic settings, but a good sustaining food supply was necessary to keep the energy machine going. The higher sensitivity and awareness of the triangular sensory receptor (eye+ear+nose=brain) gave these early mammals plenty of information and there was much room in the skull for this information to be integrated by the brain. As they worked through their unique repertoire of adaptations and expanded numerically, they probably gradually and inexorably elbowed the therapsid reptiles into oblivion.

The dinosaurs and the mammals coexisted for about ninety million years. We know that about seventy million years ago the dinosaurs suddenly disappeared. In the thirty to forty million years before their ultimate demise, they produced a number of odd-looking, presumably capricious types, typical, we should say, of creatures long into their ecological niche. Having little dramatic to say anymore after *Brontosaurus*, *Diplodocus* and *Tyrannosaurus Rex*, they now produced armored *Triceratops* with its plated back fringe, and *Stegosaurus*. Bizarre signs of senescence, paleontologists say.

Yet the mammals had coexisted with them for so long that we wonder at the cause and effect. For within five million years of the disappearance from the record of dinosaur fossils a plethora of mammal types appeared in almost every shape and size imaginable. The gate was open and the mammals from about sixty-five million years ago (the Paleocene) were off and running. What happened?

Most of the hypotheses about the extinction of the dinosaurs refer to the climatic changes caused by the upheaval of mountains—the Rockies, the Alps, the early Himalayas—or even astronomical cataclysms. Changes in climate could have destabilized the dinosaur. Could it not also be that the changes in ecology and climate caused by the mountain uplift gave the mammals more breathing space at the same time as they disturbed the quiescent, even declining domination by the dinosaurs? Would not a slightly larger and more aggressive mammal with a better brain to begin with—and now even bigger—cause the dinosaur some problem along a wholly new avenue?

The mammals would not have to attack these creatures bodily, a hopeless task in the first place. They could have attacked them where mammals have always been dangerous to reptiles—the turtle is an example—and their avian descendants, in the nest. As we can well testify, mammals love eggs—scrambled, easy-over, in omelettes. Larger, more aggressive mammals could have now added this new delicacy to their diet. It need not have taken long at all. It could have been quiet and quick.

Scientists have recently reopened an old debate. Is evolutionary change continuous or discontinuous? Does it possibly proceed by great leaps or by sudden enormous mutational changes (punctuational equilibrium)? Here in the evolving story of the crossopterygians, the therapsids, and later, of our own simian ancestors, is a hint of the probable answer. Great evolutionary leaps are rarely made by previously dominant, well-adjusted life forms. They are often made by the opportunists that are lurking in the background.

Not being ubiquitous, common, and widespread, the evidence at so many million years removed has not been found. Since the new and sudden opportunities give the obscure one a great "once-in-a-lifetime chance," it rushes out and makes good use of its store of variation. Evolution thus appears discontinuous. It does have its different tempi, sometimes lightning-like, at other times placid. Evolutionary change is wonderfully surprising at times, but it is not miraculous.

# IV

# First Crisis and the Birth of *Homo*

## Crucible of Creation

Is thirty million years too long to wait for a place in the sun? They say that all things come to those with patience. It is clear that man's ancestors needed the time to give expression to their enigmatic nature. Here we are now, a dozen or two millennia into our own. What have we achieved toward our long-term evolutionary destiny? A few brief moments of sublimity. More often either tawdy or horrific. Prognosis unknown, definitely precarious.

The original prototype was created in the Oligocene. Like the crossopterygians and the therapsid mammals, it was for a long time a creature lacking in adaptive charisma. Subject to extraordinary pressure, it almost expired. However, in that crucible of intensive mutational chaos, it developed differently from its ape and monkey cousins. Then fortune beckoned and it gradually came forth, first to devour its immediate competition, finally to change the ecological face of its earth.

A familiar story? That is why it is so plausible. Both direct and circumstantial evidence bears out our claim, first in the bones, the rocks, the artifacts, ultimately in the product. We are not so different today from the day we were born. Any mother can tell you that.

## Mammalian Dynamics

To be one variant among many is not special. However this is probably what the prototype hominid was. He was now, in the Oligocene, part of a line of mammals that was truly an evolutionary growth stock. This was not always so.

When the mammals had broken out of the reptilian encapsulation in the Paleocene sixty-five million years ago, there was an Oklahoma land rush over the face of the continents to find a secure and dominant ecological niche. Talk about speedy evolution, that was it! As in all races, not everyone got to the starting line on time or in the best position and true to the evolutionary tradition some just got elbowed out of the way.

It happened at the beginning and it happened later on in mammalian history—witness the bats, porpoises, whales, seals. Man's ancestors were lucky here. They were able to scamper up trees as others lumbered onto the road to giganticism—*Titanotheres*, the largest land mammal ever, or *Cynodictis*, an Oligocene predator. It is interesting to note that these early successes swelled in body bulk but not in brain power. So good was the living that they didn't have to use their heads to survive. (If this sounds like the story of the ancient Romans or contemporary Americans, so be it.)
Intelligence is the slowest, most tedious of structural developments. The welding together of that myriad of neurons and axons into a web of synaptic information processing, thence linked to the complicated sensory receptors, needs plenty of time.

The early mammals took ninety million dinosauric years to break away and the early primates who scampered up the trees to get out of the whitewater current of mammals needed about thirty-five million more years to put it all together. As they went up and down those trees, developing binocular vision and cleverly evasive behavior, their brains developed. First the tree shrews, then the prosimian tarsiers and lemurs eventually climbed down from the trees to a more variegated existence and then suddenly the intense pressure was gone.

In those thirty-five million years from the Paleocene to the Oligocene, the mainline mammals had dug themselves into their basic adaptive tongue-and-groove dependencies— food or fodder—that in effect made them oblivious of all else around them. The new variety of primates of the Oligocene, whether proto-anthropoids (apes, monkeys) or hominids, found a world much freer of pressure than before.

Of course they had grown larger, perhaps one to two feet tall, and heavier by ten to twenty pounds. Mostly they were smarter, with good evasive skills. That proto-ape of the Oligocene thirty-five million years ago, *Aegyptopithecus*, strikes us as a still malleable creature who could be many things in many situations. Naturally, prowling on the forest floor or climbing up trees, or both, these creatures weren't about to concede an inch. They were off on their own sweepstakes.

Fifteen million years later in the Miocene period the dryopithecine apes appeared and Proconsul, precursors of our own degenerate contemporary great apes, now bigger and definitely beginning their specialization. We are now aware of the many able but recently extinct ramapithecine apes, seemingly closer in structure to man. Crowding nearly always demands specialization, unless competitors can be eliminated in one or two strokes. This was certainly not possible in the early Miocene twenty million years B.P.

## Preadaptation: the Struggle for Survival

Where was *Homo* or even his brother-in-law *Australopithecus* then? Not to be seen. Why? Probably because the poor creature again got left behind in the rush of pongid expansionary development and specialization. You just don't get left behind to brood about your ill fortune. While no creature has control over its own destiny—it cannot act on its own behalf—nature's selection process is not so reticent.

As in the case of the early mammals, the outside pressure elicited a counterpush from the inside. One can see it as the mutational dice working overtime, if not in exuberant expansion, at least in desperate defense. A set of accidental features—vocalization, brain and hip girdle structure, ambulatory habits—along with minor yet characteristic elements of penis, hand, foot, and face were articulated in a unique animal, all through a frenzy of mutational changes. The basic cause was the hyperselective conditions under which these creatures must have lived.

Literally from generation to generation, the early hominids were honed by survival or extinction to the demands of the pongid expansion. No tigers, lions, or elephants did they have to fear, off in the distance, merely the elbowing pressure of pongid cousins. If there had been long-distance selective pressures by other animals, our biological responses would be far more physical than they are. In reality we have no good physical defenses against such beasts. (That is what made the Roman circuses such exciting entertainment.) It was no contest until man obtained a weapon other than his body.

We can never know how our ancestors hid out, what overt means they took to defend their right of existence, but since they were like many other furtive mammals, we can imagine a scenario. First they stayed small and unobtrusive. It was of course not a matter of choice, simply the fact that smaller specimens could slip into a crotch of a tree and be camouflaged or could do with a little less food and still inseminate their females. These are typical adaptations of defensive creatures.

More revolutionary was the selective directing of this creature's unique structural and behavioral characteristics, made necessary to its survival. The special vocal structure, the peculiar brain, and perhaps even the accentuated intelligence of evasion were now being melded into a powerful alloy, a strategy of survival. Not simply a melange of screams and hoots, whoops and sounds indicating a variety of emotions, needs, and information, as with the apes. The brain may have been highly enough evolved as time went on to perfect a new communication system, a shaping of sounds to stand for a lexicon of words, now helping to clarify information and messages between members of the family.

The bonding of female to dependent child and the widening range of sexual dimorphism for behavior here came into play as powerful welding elements. The child needed the mother. Being more intelligent and bigger-brained, it was probably more instinctless and helpless at birth. The mother needed the father to provide food and protection. Here a more or less monogamous commitment, including continuous sexual receptivity, was forged in the precarious defense of the maturation and survival of the next generation.

Unquestionably the social bond among humans is powerful. Can you doubt that the nuclear family, extended or not, was the key to the defense of the individual? This is true even in our own casual times. The nurturing and the training of the child, as symbolic intelligence replaced instinctual signals, had to be long extended in time. Language became a powerful adjunct of the social glue. Again, as humans were gradually freed of the limbic call system by the expanding cerebral cortex, language became a vehicle of discourse, primitive at first, but far more explicit and powerful in its informational content than the communication system of the pongids.

Language brought more dividends. As language grew from this spontaneous ability to vocalize literally millions of discrete sounds, each hominid group could create its own vocal repertoire. Robins may have song dialects at different points in the forest, but these are instinctual. Human groups freely created different languages. Even a human with an I.Q. of less than forty (a microcephalic idiot) can speak. Not much content, yet the language structure is fully hominized. Powerful human centripetal feelings and relations get caught up in language. For the earliest hominids, language was used both to communicate survival information and to keep the group together, to bind the group emotionally as well as practically, more powerfully even than could its now departing instinctual repertoire. Here was a uniquely flexible behavioral tool.

One wonders how all this was coordinated, how such a transformation could have occurred, how mutations could so coincidentally present themselves for use. In fact, a new systematic reconstruction seems to have tied all these diverse strands into that uneasy amalgam of traits we call man. The concept is now well established in biological science. It is what has been called "paedomorphism." First put forth in the 1920's by W. Garstang and brought to elegant fruition by Sir Gavin de Beer, this concept attempts to explain a wide variety of evolutionary restructurings in animals that often have significant positive results.

Paedomorphism is the process by which immature characteristics of the ancestral type are retained by the mature descendant. Human nakedness is an easy example. There are others: neotony, the sexual maturation of the biologically immature, and caenogenesis, the slowing up of maturity, both related to the fundamental work of the rate genes. In man, the most interesting change, an example of paedomorphism, is the manner in which the location of the skull opening for the spinal cord is in the same position as in the foetus of other mammals. In mammals, this angle is gradually changed as birth nears, so that, finally, the opening is nearer the rear of the skull.

Man's foreshortened face is reflected in the foetus of most mammals. Indeed the apes look very human shortly after birth, but soon develop the typically prognathous profile, the great bony ridges on the skull, and enormous jaws. Humans are baby-like into maturity and old age. These facts should tend to confirm the conservative nature of man's progress, his combination of ancient and modern physical characteristics.

We think that paedomorphism has important adaptive uses for a creature undergoing powerful external selective pressures. For example, if the animal could retain its immature body and behavior, it might also retain the obscurity with which natural selection often endows the immature—perhaps like the spots of young lions. A. C. Hardy has called this process covert evolution, "escape from specialization." Here is a possible explanation for the origin of our vocal and language system, its odd bicameral brain specialization. All these characteristics of the past, of the childhood of the same creature, were now, in one great mutational revolution, catapulted into maturity.

## One Step Back

Think of the vast proliferation of mystery religions during the imperial era of Rome. Who could have picked Athanasian Christianity as the winner? What a rich and odd combination of rituals, gods, saints, and mysteries these religions contained! What would be the triumphal combination and how would it win? In retrospect we know that Christianity was the natural solution. In retrospect, also, we now see *Homo* as the light of the future. The many almosts—hominids, great apes, ramapithecids, even monkeys—who ventured onto the plains, however, testify to the experiment of the genes, the vagaries of nature.

For our first true hominid, the pongid revolution was an opportunity to hold back, to take those lonely steps of restructuring in tune with the needs and inevitabilities of a greater brain and a richer, wider intelligence. Mankind came forth with a body to assist the adaptive work of the brain.

Just as importantly, and mysteriously, a social structure was created that could insure his genetic immortality.

# V

# Breakout

## Vault into Success

"Apes can brachiate faster than trees can retreat." So chuckled Harvard anthropologist Ernest Hooton in the 1940's after listening to some of the explanations of why man turned his back on his tree-living relatives and took to the ground and the bipedal upright pattern of life.

The climatic changes brought about during the Miocene, twenty-five to seven million years ago, by the upthrust of the new mountain ranges—the Andes, the Himalayas—were to be highly significant. Large areas around the globe were gradually opened up and grassland, plains, and highland ecologies supplanted much of the uniform dark green to purple mantle. During this period, the hominids began to make their move. With them came a wide assortment of hominid-like monkeys and apes. The outside world was changing. Some would be readier than others.

We presume that at this point, after perhaps fifteen to twenty million years of encapsulation, our primeval ancestors were ready. The conditions of jungle competition with fellow primates must have divided the protohominids into a series of near-related types. It probably forced a variety of similarly positioned creatures (*Oreopithecus, Gigantopithecus, Ramapithecus*) into that same genetic-environmental-selective dynamic that produced those changes in man of which I spoke in the last chapter.

Consider that *Ramapithecus* had a foreshortened face and thickly-enameled, human-like teeth, that the Italian Miocene ape *Oreopithecus* also had a foreshortened face, human-like pelvic structure, that *Australopithecus*, our closest ape-man relative, had a Broca's area in his brain and presumably some language ability, that *Gigantopithecus* from India, again with a human-like facial structure and a ground-living, open-country lifestyle, existed for so many millions of years in those Himalayan foothills. All this argues for a wide variety of creatures partially preadapted to the new ecological circumstances and showing the effects in their various physical attributes of the competition and variation in the Oligocene and early Miocene jungles.

We have for so long, and unfortunately almost universally, attributed everything that is unique to man as having a direct, concrete and explainable, practical, or survivalistic element. If man stands erect, it is so he can more easily look over the high grasses for possible enemies. If he engages in frontal—face-to-face—coitus (so unlike the apes) it is to effect a more intimate relationship with his female, thence to bond the family. So it goes almost *ad infinitum*.

Look at man today. Ask yourself, cannot anything that man does successfully have at the same time a negative selective result for his survival and prospering, e.g., reproduction, religion? Someone once remarked, with truth: there is no stimulus, no temptation, no need, to which man *cannot* say "no." The fact that he is as free as he is from the specific "drive reduction" behavior to which all other animals are subject ought to tell us something.

Man is a creature of biological evolution. He is, however, neither ape nor monkey, but something very different, even if related at a distance now of some thirty to thirty-five million years. A lot of water ran under that evolutionary bridge.

Success is a great thing, but it is also dangerous. It leads toward great morphological, physical, as well as behavioral opportunities. Just as in our own time, success brings with it a host of human obligations; it also brings the hangers-on and parasites who limit one's freedom. So too, evolutionary success, getting good at something, always leads to opportunistic specialization. Then, when the wheel of time and fortune spins, the successful of this era become the redundant atavists; in the next era they become over-specialized, obsolete, then, finally, extinct.

Until late, the characteristic of evolutionary success could not have been used to describe the hominids. They hung back, forcibly or otherwise, and waited. Only in the Miocene did their repertoire of preadaptations come into use. The key element was their *lack* of specialization except for that superior intelligence that allowed them first to escape extinction in the Oligocene, then to turn their existing repertoire of tricks to efficient use within a set of new environmental possibilities.

**Wanderers on the Plains**

To try to answer that original question, why wouldn't an ape brachiate faster than the trees could retreat, we must explain as follows: By the end of the Miocene, the mammals had been in the expansionist phase for at least forty million years, the anthropoids about half that time. Life itself stretched back over 750 million years. Think of the repertoire of experiences that had been recorded in the genetic time machine carried in every body.

The basic principles of adventure, experiment, new possibilities had been branded into the very nature of animal life. For it is in these primal amino acids that make up the structure of heterotrophic life (life that cannibalizes organic matter—its own fellow creatures), that at the beginning opted for the journey of the "variable cuisine." While the nonorganic energy-food consuming autotrophs took their chances with the quiescent environment—clays or the sun— the animals went on their journey with history and time searching out living food wherever it could be found. Instead of roots, they grew hairy tentacles, they grew fins and wings, they grew feet that marched over the earth in a great drama against energetic degradation: entropy.

Thus forevermore, while so many of their ilk— paramecia, snails, worms, barnacles—settled into comfortable niches, others marched on. It was in their blood. It is in ours. As we all know, for every ten people who stay put comfortably, three or four pack their bags, give up everything, and take their chances with destiny. Ten years later, of these four, perhaps only one will be better off.

We are all wanderers. We ought to be able to calculate however, our chances of landing on our feet. That is the difference between the winners and the losers. Every time I enter Jack August's Sea Food Restaurant to order a boiled lobster, I think of that half-billion-year separation between the arthropods and me. So secure for so long, yet I am squeezing the lemon juice on the lobster, not *vice versa*. You, skeptic, will laugh, for outside the screen door, hurtling toward the light are other arthropod cousins. Will their day come? Our smiles may fade uneasily at the thought.

The Miocene was a rough era, for it was the time that all the modern forms of mammalian life were shaped. The faces—horse, elephant, tiger, hippo—all look familiar now. Many did not survive into the recent era and the attrition rate from that point on is alarming. Today, it is virtually extermination.

An interesting and important fact: In the Miocene, intelligence in the mammals became an important desideratum of survival. Having shown all their brute physical tricks, having filled up the vacuum, even the seas and air, the mammals had to make do with less room, but they needed more savvy. The whales, as soon as they tumbled into the crowded oceans, found that out—too many hyperactive teleosts to compete with. Within fifteen million years they had achieved maximum brain size (it occurred in the Miocene) and since then they have remained in stasis.

There was only one area of opportunity; that was on the savannahs of Africa, Asia, and Europe and then only for a featherless biped with some smarts. Here a new, at least more overt, evolutionary drama played itself out for man. The protagonists are more familiar. They were probably our collateral ancestors. As the Miocene (twenty-five to eight million years ago) slipped into the Pliocene (seven to two million years ago) they appeared on our stage and quietly exited. *Homo* took his time; he awaited the last act before he left his bony calling card some three million years ago.

## Cast of Characters

Let's review the personages in this second act of hominid evolution: (1) *Ramapithecus*: ramapithecine fossils, first discovered in the 1920's, are widely distributed in Africa, India, and Europe. *Ramapithecus* dated from the end of the Miocene, fifteen to eight million years ago. A little creature, he was perhaps only two feet tall. Only a few teeth, facial bones, more recently fragments of the rest of his body (postcranial bones) have been found—short, deep face, very unpongid teeth (short and vertical, the cutting edge of which seemed to have been continuous, thickly-enameled as are human teeth). It is now doubtful that he was a hominid, despite the similarities. More probably, on the basis of the recent fossil evidence (David Pilbeam) as well as his ancient widespread distribution, he was at most a collateral ancestor (not in the direct line).

(2) *Australopithecus*: a controversial story here. Not too long ago, most, but not all, anthropologists hypothesized two distinct lines of these creatures as making their debut about five million years ago. The smaller of the two, *A.*

*africanus* (two to three million years ago), was delicate, about three feet tall, with an endocranial (inside the skull) capacity of about 450 cm$^3$. (A three-foot pigmy chimpanzee had or has an endocranial capacity of about 300 cm$^3$.) With *A. africanus* there is a jump of about fifty percent relative to the body-to-brain weight of *Ramapithecus*. (*A. africanus* was a forty- to seventy-pound creature.) Remember, however, this advance took place over eight to ten million years.

A restudy of the cranium of the now ancient *A. afarensis* (of *Lucy*, Donald Johanson fame) shows that it was extremely primitive, even when compared with a typical ape crown (Ralph Holloway). Owen Lovejoy has argued that this creature's pelvis at about four million years ago would have delivered only a very small-brained infant. Russell Tuttle of the University of Chicago believes that *Lucy's* "pelvis and feet still retained features of tree dwellers, implying a recent transition to terrestrial life" (*New York Times* August 3, 1982). Contemporary with *Lucy* the so-called Laetoli hominid at three-and-a-half million years ago, had a different footprint, essentially modern. Warning: *A. africanus* is a controversial creature, probably not ancestral to *Homo*. The ball is still in the air. Richard Leakey claims that *Lucy* is a composite skeleton made up of a number of individuals.

The other and more recent line of *Australopithecus* was the robust type that evolved more recently and lived on to a time when *A. africanus* had long disappeared. Here were larger creatures, some as big as five feet tall, with huge jaws and a sagittal crest (a ridge of bone along the midline of the skull to support the jaws). Obviously vegetarian, as compared with *A. africanus*, who had a more omnivorous diet, the robust australopithecines seem to have made their last stand in the jungles alongside that other more traditional and retreating host of contemporary apes.

It didn't help them. Not quite fully evolved bipeds (this

issue is also contested—some say that they were fully bipedal, bone structure notwithstanding, like man, M. H. Wolpoff), the australopithecines are presumed to have walked in a rocking, rolling gait, certainly not good enough for a long-distance march with the United States Army but probably functional in the mixed forest, plains ecology where they hung out.

Casts by Ralph Holloway of the inside of fossils' skulls of the robust *Australopithecus* reveal markings of a Broca's area (language) and from this we assume that australopithecines had the rudimentary language skills that define humanity. It is in the story of their tenure on earth that we get a first sense of the human dilemma, a message about man, even while his first appearance is still screened from direct view. For if we are not quite positive about the evolutionary position of ancient *Ramapithecus*, we are more positive about *Australopithecus*. Sure, anthropologists like the Leakeys argue one way, M. Taieb and Donald Johanson argue another about who discovered the earliest *Australopithecus*, whether it should be called "*afarensis*" or "*zinjanthropus*," whether it was one large superspecies or a series of related species of one large genus. It is still a matter of contention whether *A. afarensis* (*Lucy*) was on the road to *Homo*. We can disregard the anthropologists' squabblings for scientific fame and fortune as predictable, as a hoary part of our scientific or cultural race for honor and its material emoluments.

What is clear is that the australopithecines were our neighbors on the plains of Africa, and then we lost them. We may have never seen or bumped into the ramapithecines. The australopithecines, however, we met, and probably destroyed.

(3) Richard Leakey first claimed that "1470" man, *Homo habilis* (the first-named truly human ancestor of *Homo sapiens*) goes back almost three million years. This has now been attacked by various anthropologists, obviously of the

opposing camp (they claim it is only 1.75 million years old).
Dating these fossils is a problem for many reasons—the need
to interpret the strata of rock in which they are embedded, the
possibility that the fossils have been found in disturbed
strata (which could be misleading), as well as the margin for
error in the dating techniques (potassium-argon, carbon 14).

Needless to say "1470" man was an old, very large-
brained *Homo habilis* and certainly within the time
framework of the australopithecine tenure on earth. The
possibility is that it is a female skull. Thus its approximately
800 cm$^3$ capacity, on a body speculated to be about four feet is
extremely impressive, as female skulls are smaller on
average. The usual *Homo habilis* ranged around 600 cm$^3$.
Hence this was a much brainier creature than
*Australopithecus* pound for pound and inch for inch. We
measure the relative intelligence of mammals by their
endocranial capacity relative to height (not weight).

About one-and-a-half to one million years ago, the
gracile *A. africanus* disappeared. At 750,000 to 500,000 years
ago, the forest-living, robust herbivore *A. boisei* disappeared, having outlasted his brother *A. africanus* by a
considerable number of generations. Why did these creatures
die out? Weren't the australopithecines as capable of
surviving as our more distant relatives the chimpanzee, the
gorilla, the orangutan, even the free-swinging gibbon? Until
only a century or so ago, these more primitive creatures had
the jungles of the land continents of Africa and Asia securely
to themselves.

# VI

# The Painful Truth

## An Origin to Aggression?

I wonder at the difference. We are anthropoids, primates. Travel back mentally to the early tarsiers and lemurs, to the tree shrews. Furtive, hidden creatures, displaying timidity in every glance, hesitant. Are these our progenitors?

Today the evolutionists are aflutter with puzzlement and consternation. With each new discovery of a fossil fragment, the pedigree of man and ape seems to become more baffling. For awhile, David Pilbeam persuaded us that gentle little *Ramapithecus* could be a candidate for *Ur*-hominid status. Delicate, thickly-enameled teeth, seemingly a paedomorphic bone structure, all augured possibilities. Then came a few postcranial bones. Definitely not a biped and the whole theory evaporated.

Interestingly, more bones of the ramapithecids have now been found. The family that includes that great but placid vegetarian, *Gigantopithecus* of the Himalayan foothills, was ubiquitous in the Miocene. About eight million years ago, with the exception of *Gigantopithecus* and one lone tropical descendant, it disappeared. That descendant was our own orangutan, also with thickly-enameled teeth. Yet he is so different in his brachiating habits from our envisionment of this ancient line. Has our orangutan regressed, was he pushed into his isolated Sumatran sanctuaries by mor aggressive invaders? Were these same unknown intruder responsible for the demise of other members of this line?

If the orangutan is explicitly quiet and forebearing, then it is a greater surprise to find that his omnivorous African cousin, the gorilla, is likewise gentle, if not the soul of extroversion. Nowhere, even among the aggressive ground living baboons, do we find the blatant antagonistic commitment of *Homo*. How do we explain this shift in the external adaptive stance of the primates? Nowhere in the fossilized record before man do we find any clue to indicate that we are not alone. We are a sole, ferocious, aggressor anthropoid.

There is no question that both chimpanzee and baboon will sup on a dinner of meat if the opportunity avails. It is almost like the casual dinner of human flesh that a tiger or lion will enjoy if the circumstances arise. The simple minded intention to kill is not there. Can you doubt the killer in *Homo*? In the casual shufflings for living space, to obtain an inch on the jungle floor, our ancestors were pressed to become *Homo*. Still, these defensive adaptations gave no hint of the future. When were they transformed? How and why did we come by that special if problematic human uniqueness: war, aggression?

## Early Genocide

Let us go forward in time from the Miocene jungle womb. A book, *Life*'s "Nature Library," *Early Man*. A colored drawing. On the left a group of bipedal creatures with many ape-like characteristics. They are mostly turned in flight—males, females, young obviously in terror, a few large males hanging back to throw rocks and other missiles at some intruders.

Whom are they fleeing? The enemy appears to be slightly smaller in size, far more human-like in face and body. There re only males, and they are running forward in almost army ormation, exhibiting ferocity of face and body as they rigorously shower the retreating australopithecines with tones and rods. The enemy is *Homo habilis*, "1470" man. The retreating australopithecines are the robust kind, those eft behind after the disappearance of the gracile *A. fricanus*. Now they are in retreat, escaping from the open plains to the nurturing protection of the jungles, back to heir ancient but safe diet of fruits and berries.

It would take a while longer before *Homo* would again eenter the jungle. Then he would have a different brain, lifferent intellectual and physical armaments. He would not bother to go after the evasive and foolish chimps: he would merely push the gorillas aside and exterminate the last of hese australopithecines.

Why? Because little guys, put upon when they are young annot help playing the bully when they are older and bigger. Simplistic, you will reply, and, of course, you are ight. Nevertheless, we would have to be utopian to ignore he killer in us now. There is absolutely no logic in viewing man as living in a Peaceable Kingdom at the beginning of his generic journey into the present.

The fossils hint that man was the brainier, less agile, probably smaller creature at the time of the Miocene climatic transformation. While the australopithecines ambled awkwardly, but with instinctual assurance, onto the savannahs, man had to struggle with dependent young, highly pregnable and not easily mobile females, and the chaos of excess neurological excitation. He was not yet up to creating an efficient social means of coordinating life within he group or gaining a living outside.

This took time, perhaps six or seven million years. Gradually he learned to cope with life on the plains, avoided he forest apes and the open-country australopithecines, also he dogs and cats. As he wandered to obtain his living, he grew in size, and his brain grew allometrically (hand-inglove with the body), or perhaps even more. At a crucial point of body and brain size growth, the defensive protective net of family and language now became his opportunity. H had learned to secure the hearth and now went in search o prey.

The Pliocene (seven to two million years ago) was a transitional period. During it the hominids were presumably making their evolutionary move forward. At the same time, the number of ape species inexplicably shrank. With this shrinkage of apes in both variety and number, the monkeys came into their own.

The reader knows what I am leading up to. Even from the time of *Australopithecus*, *Homo* has been a killer and a cannibal. In his book *Origins*, Richard Leakey envisages a mythical Garden of Eden in eastern Africa on Lake Victoria two to three million years ago. Here *Australopithecus* and *Homo* and hundreds of other species passed each other with a nod of the head as they were smothered in honey and cake by the rich ecological environment. Not likely!

Wishful thinking, a joke, hardly human nature in this picture, hardly nature's fate. In the beginning, the good life beckoned, Leakey claimed. Then came the hordes, the honey turned to gruel. The masses, only then, aimed for the jugular to secure a bit more than the remaining crumbs.

We can now observe in many fossil skulls the torn-out foramen magnum (where the spinal cord enters the skull). The brain was good eating and, of course, good protein. The bones contained marrow, and we do find them split and drawn. Then too, perhaps this was part of ritual. If *Homo* didn't decimate the versatile apes of the late Miocene and Pliocene, then *Australopithecus* helped. Thence, he got a dose of his own medicine.

When the lands emptied, down from the trees came another fellow, the baboon. Plenty of room on the ground for someone who was different. He is there today, filling in where once were versatile apes and hominids, intelligent capable cousins who could have given us a glimpse of our past and perhaps a more humble perspective on our own nature and its dangers to life, including *Homo sapiens* himself.

## Why the Killer?

The question is "why?" How do we explain the transformation of a creature whose basic adaptive structure in the beginning was defensive to one that increasingly over time became genocidal? The problem of human aggression and war has tantalized all thinkers and many have plunged in to try to explain it. Few have been convincing. I am not going to go into a detailed and elaborate exposition here, merely to underline briefly what I think were the conditions and circumstances of this transformation from a defensive, evasive anthropoid to a deadly, aggressive one.

Basically, I think the issue takes shape at that point when man began to rely more on his social intelligence to solve the problems of his life than on his instinctual, stimulus-response system. Man has few biological limits on almost anything he does, only those of intelligent prudence or thoughtful consideration for possible effects. Often the less thoughtful our responses or the more gut reactive, the more we show a mammalian set of limbic system actions and emotions of sometimes terrifying ferocity, but without those natural restraining elements that nature builds into the instinctual behavior of animals.

As man turned onto the plains, he had to take what he had in the way of morphology and turn it to use for survival. Unlike the horse, he couldn't grow hooves, but he obviously put his best foot forward. In this case it was a highly developed brain. Like the mammals of millions of years earlier, a little bit of growth and that extra bit of brain power made the difference. What grew especially were the cortical, evaluative areas. Those skills that were necessary earlier for

the defense of the family would now, in a more opportun
environment, gain in effectiveness. Because of the openness
of the plains, distances for finding food were longer, whic
almost demanded increased size. As the cortex expanded ma
was pushed onto a new level of adaptation.

We know of the relationship between a carnivorous di
and the needs of the brain. The brain demands tremendou
amounts of highly oxygenated blood. The greatest source c
such a rich blood supply is animal protein. A kind of servo
mechanical feedback arrangement must have existe
between a growing creature in an open ecology with fewe
vegetable resources and the richer if distant animal food. Hi
brain, now thriving on meat, in turn grew to emancipat
man from his timid diencephalon (limbic system) respons
system of the Oligocene and early-to-mid-Miocene origin:
The result was the channeling of enormous nervous energie
through cortical thought.

As is the case now, early man presumably coul
voluntarily turn his ferocity on and off. To his family he wa
a devoted explorer and protector, sacrificing the fun an
games of pongid jungle life. The rewards of his emigratio
onto the plains were clear. Instead of being enclosed in
protective jungle, always reacting to threatening outsid
forces, now his energies were expansively open and exciting
The new or possible threats didn't hold nearly the terror a
the old. Man never needed a finely honed instinctive surviva
system. Nor did he need to develop highly articulate
physical adaptations to survive. No claws, no fangs, not ever
great speed or size were developed.

Man's aggressiveness, his killer habits seem to hav
arisen as a consequence of the above transition from on
phase of life to another, from one level of behavior to
completely new one. Instinctual inhibitions no longe
operated. The symbol system of cultural meaning deter
mined his attitudes and expectations. Rooted in family an
band, ethnically defined in language, tradition, ritua

worship, man saw the world outside now as strange and meaningfully threatening. Man was both fearless—he had no natural enemies (except for himself)—and fearful. These fears were now only in his mind: "they" and "we."

Somehow the growth of the cortex had been accompanied by the generation of powerful mental and emotional excitations. The passions burst upward from the mammalian limbic areas of the brain through the symbol system organized in the cerebral cortex. *Homo* had to sort it all out.

Man could not be expected to understand the significance of these new forces immediately. In the beginning he fell victim to the simplest "we" and "they" triggers to action and destruction. Eventually his growing brain and his maturing sophistication would even dissolve those basic biological distinctions and he would be culturally able to go for the jugular of even his closest kin.

# VII

# Homo Stabilis: the Model of Man

## The Hominids Move Ahead

The Pliocene was a fruitful period. At its beginning, some seven million years ago, it probably had a greater variety of mammals than any other period. Practically all the recent forms of mammals could have been recognized, in transit, as well as innumerable extinct varieties, either passing into oblivion or giving reasonable signs of so doing. For the five million years that the Pliocene covered, the pace of evolutionary change quickened in the mammalian class, reaching a frenzied conclusion at its close with the beginning of the Ice Ages, the Pleistocene.

Intelligence and specialization were the key elements in this calculus. Mammalian intelligence, in the struggle to survive, was honed to its finest edge. Eventually, however, the tongue-and-groove dependencies such as predator-prey created in its competitive honing the thread of ultimate decline. For in the background the generalists were gaining strength. Catapulted above and beyond this adaptive fray, the hominids, first in a trickle, then in a rush, were added to this moving equation.

By the middle of the Pliocene, the australopithecines must have assumed their basic form. Soon after, the gradual specialization of the species and then even of its component aces showed itself in a wide variety of forms, from the early *A. afarensis* to *A. africanus* to the various robust vegetarians that eventually slipped back into the jungles.

It appears that the earliest established date for a true human, *Homo habilis*, "1470" man, is close to the beginning of the Pleistocene, 1.8 to two million years ago. Richard Leakey would have him a million or so years older. Nevertheless the probability is strong, as noted earlier, that *Homo*'s Pliocene history is independent of the australopithecine man-apes.

By one million B.P., a very short time even if we use the brief timetable for *Homo*, *Homo* had advanced in actual physical size from 3.5 feet to almost five feet. He had also begun to make the transition from *Homo habilis* to *Homo erectus*. In brain size (endocranial capacity), he had gone from an average of about 600 cm$^3$ ("1470" was a remarkable 800 cm$^3$) to 800-900 cm$^3$.

*A. africanus* was on its way out, if not already extinct. The closely related great apes of the late Miocene had all but disappeared, with the exception of the specialized and regressive contemporaries we now sport in our zoos. A few more hundred thousand years and the last of the vegetarian, big-toothed, forest-living, robust australopithecines had disappeared. There was no future for them, no transition from one successor species to another. In fact, with the coming of *Homo erectus*, the biological link to the past had been effectively cut. Man had cleaned the canvas before taking on more distantly related animals for his peculiar aggressive and fatal games.

What kind of creature was *Homo erectus*, what did he represent in this ongoing journey into the future? For one thing, he was the creature that should have remained as the model of man. He was the successor to all the developing hominids. Of course *Australopithecus* and *Homo habilis* did much of the dirty work, but *Homo erectus* finished them off.

*Homo erectus* was a wanderer; he traveled long distance and with the exception of North and South America definitely populated the continents. He developed a culture the Acheulean, which lasted about one million years from it earliest stages about 1.2 million B.P. His hunting prowess was remarkable. In Torralba, in Spain, two hundred kilometers north of Madrid, a vast graveyard of elephant was found—dating from 300,000-400,000 B.P., admittedly a the latter end of his tenure—that gives testimony to an erectine hunt and feast (F. Clark Howell). Yes, *H. erectus* had discovered the uses of fire. Whether he was able to make a fire at will is another question. In short, he had little to fear.

Most controversial of all, and this is just an educated guess, he was probably even more closely related to us than his taxonomic grade would indicate. It is my hunch and I suspect that anthropologist Carleton Coon would agree that *Homo sapiens* and *Homo erectus* would be interfertile My supposition is based on the fact that over time, as the mammals wandered over greater distances, interfertility was more and more selected for. The different forms of contemporary sapiens are all easily interfertile Hypothetically, the differences in brain and bone are not so critical to have caused the genetic patterning to be blocked, which would have led to intersterility. In fact, I wonder, more tentatively here, whether the interfertility might even extend as far back as the "1470" *Homo habilis*. I, for one, would not be surprised.

In the spring of 1984, a great exhibition of the actual fossil skulls of our predecessors was being convened at the American Museum of Natural History in New York City. At the conference that preceded the exhibition (April, 1984), much discussion centered on the reclassification of these fossil hominids. The appellation *Homo sapiens* is now being hung on ever more ancient representatives of our past. Will we soon absorb *Homo erectus* into our species?

**More Related Than We Would Admit**

Over one hundred years ago, Alfred Wallace, the friend and rival of Charles Darwin, wondered why nature, having created fossil man, with an endocranial capacity of 900 cm³, (Wallace was acquainted with some of the early discoveries) would go on and outdo itself by creating a creature such as *Homo sapiens* whose endocranial capacity could reach beyond 1600 cm³. It is a good question, for in truth *Homo erectus* was truly the fulfillment of the hominid progression.

Having seen some replicas in museums, the reader may protest, "not that brutish monster." True, *Homo erectus* had none of the delicacy of *sapiens* and seems a far departure from the gracile monkey-like paedomorph that we started out as on our way out of the garden of the apes. Heavy brow ridges, massive jaws, and uniformly thick bones give one a sense of rugged stability, albeit on the scale of a five-foot creature. One looks at the skull from the front and sees the hip-roof modeling of a small ranch home. Not much swell or majesty. Yet *Homo erectus*, wherever he is found, was indeed a competent human.

He did not invent culture. *Homo habilis* could speak, work in groups, had a clearly defined technology and roughly fabricated tools with which to perform a small repertoire of tasks. The gradual transformation of *habilis* into *erectus* was accompanied by a qualitative surge forward in technology, reflecting the bigger brain. However, now the tools took on a new dimension.

Here was a creature, living one million to 500,000 years ago, with a brain almost fifty per cent of the size of the *most* hominized brain in our world, eighty-five per cent of the average, and he was an artist. Harry Shapiro of the Museum of Natural History in New York City has attempted to reconstruct the life of the northeast Asian version of *Homo erectus*, Peking man. It is probably as good as any description of these fellows, wherever they lived.

Their daily routine of picking berries, catching small animals, gathering shellfish when they could, tending a fire, was all directed toward keeping the family alive. Perhaps he glanced into the distance, wondering at the sky, the purple hills that met the horizon. Yet his mind could not yet fathom it all. He had a lurking sensitivity, but without enough power to break through and put it all together. So much more aware than the ape, he was coordinated in behavior to take on the awesome responsibilities of raising fragile and slow-developing infants, and providing for a wife, perhaps even several. Life expectancy for women was probably at best little more than thirty years, what with a continual series of danger-filled pregnancies. Yet, as stated above, inside what seems such a gruff exterior was a glimmer of tomorrow, of a bigness of soul, a sensitivity to form and matter, the beginnings of a mind that would one day explode.

We call *Homo erectus'* culture *Acheulean*. The overwhelming representation of his work that has come down is the hand ax. In various shapes—triangular, oval, teardrop—these monster creations, sometimes almost a foot long and seven or eight inches at their largest width, sometimes two inches thick, were often great tanks of tools.

From the earliest Abbevillian examples to the most refined creations later in erectine development there was an indescribable unity of concept. While the same tradition continued for several hundred thousand years, there was development. Meticulously chipped, often from the most beautiful marble and granite, each of these axes must have taken many days to produce. They manifest unbelievable patience, planning as well as insight into form, unity of design, beauty of faceting, and often great care in the choice of materials to enhance the visual impact.

There it was, at a million years removed, a high development of the esthetic sense. We don't know whether these erectines sang, danced in ritual, or made other beautiful objects out of less durable materials. There could not have been many erectines in any one group because the problem of food supply cut down the number that could range through any area. Thus perhaps instead of or along with the usual group rituals or ceremonies, performed by most human groups to work off their energies and exercise their esthetic delights, *Homo erectus* spent many a lonely day chipping away at various-sized rocks to create tools that often couldn't have been useful. Some were just too big, heavy and clumsy to be used against man or beast, unless the prey was practically immobile.

Sometime between one million and seven hundred and fifty thousand years ago some erectine groups drifted from Africa into Europe, western and northeastern Asia, and on into southeastern Asia and the islands of what are now the East Indies. What is puzzling about this migration is that the Acheulean hand ax can be found in Africa, Europe, and western Asia (India), but never in the northeastern Asiatic quandrant where Peking man was discovered. Why? Could it be that the Acheulean hand ax was developed after the group that eventually settled in China had left their African homeland?

# PART 2

*Sapiens:*
the cauldron of creation

# VIII

## The Brain Just Grew

**The Myth of Practicality**

Why did the brain grow? The answer is simple: more human parents with brains of larger endocranial capacity had children than the less able. Beyond that seemingly innocuous equation are controversial questions and more complicated issues. I do think that they can be dealt with both simply and persuasively. In the process we will be able to discuss one of the central aspects of what we humans are, indeed the meaning of our long, almost inevitable journey into the present.

There is a kind of "debunkable" myth in evolutionary theory that leads people to say that all biological functions, because they have survived, must have a practical reason. Isn't it obvious that giraffes have long necks to get at succulent leaves and branches high in the trees? Hooves help horses gallop over hard grassy plains, wings serve the purposes of flying. To the extent that they have brought creatures into the present, all these characteristics have been more successful in the sweepstakes of life than those that other creatures brought with them.

Thus would it not also be true that the various functions of human intelligence serve clearly describable and possibly observable survivalistic uses—speech, bipedality, hairlessness, frontal position of the sex organs, and finally, that bulging brain? Throughout decades-long discussions of the mystery of human evolution, innumerable articles and books have dealt with the practical functions and purposes of every last iota of human behavior and structure. How have they saved man from extinction, how did they propel him into contemporary dominance?

There is no question that something about man, his combination of defensive anthropoid sensibilities and intelligence along with a late-coming social aggressiveness, made him dominant. What I want to do here is to show that the equipment that man received through mutation, adaptation, and natural selection did not come about because need generated the appropriate organic responses. That view, a discredited commonsense utilitarianism, has been generally rejected. Such a manner of looking at this process of evolution in general and human evolution in particular produces the kind of bizarre falsification of human nature that is found in the behaviorism of B. F. Skinner or the ethology of Konrad Lorenz.

By now we all know that these kinds of writing caricature humans and human behavior. The only reason that such reductive pictures of *Homo sapiens* as aggressive ape or clever rat still endure is because of claims that they derive from the supposed selective practicalities of evolutionary events.

Man survived and became *Homo sapiens*; but he did not survive by behaving like a rat. The question remains, if we didn't get selected out by old Mother Nature, if we didn't destroy the dryopithecine apes, wonderful *Gigantopithecus*, our first cousins, the australopithecines, by some nonpractical means, then how else?

What I merely want to emphasize is that if we killed in order to live, we did so in a human way, just as we do today. What I intend to argue for is the method of our adaptation and survival as humans, in which we remain true to our special nature. Indeed the road to selective victory led us to perfect our nature as humans, not as poor copies of chimpanzees or wolves.

## The Inevitability of Intelligence

Central to this picture of human evolution is the idea that man's coming appears almost predestined. No, I am not getting theological here; I am not postulating a preordained journey through time for this creature now endowed with a "supernatural soul." Man's uniqueness is rooted in nature; it is not a gift from beyond.

Simply, *Homo* is nature's intelligent animal. Intelligence, one of the great adaptive modalities of evolution—as are fecundity, miniaturization, giganticism, aerial, aquatic, and land adaptations, instinct, stability, camouflage, sexuality, in other forms of life—found its most efficient receptacle in man. Man certainly is not a perfect embodiment of this characteristic. As we well know, human nature demonstrates a complex, almost schizophrenic pull between tendencies. Intelligence as the slowest and most complex of the adaptations of life required a long series of accidental events to shape a creature that could utilize it as a technique of adaptation.

The gift that we see in Michelangelo's Sistine Chapel, as God touches Adam, could be that spark of intelligence. The evidence for man's special nature is there for us all to see. Explaining the origin of the human cultural panorama is another thing. How did we acquire religion, the arts, human love, self-sacrifice for the sake of mankind, patriotism, celibacy, genocide?

They all came as part of the development and growth of that special morphological protuberance, our hypertrophic brain. Does this brain growth therefore mean that those various traits of culture noted in the last paragraph serve some survivalistic purpose? The seeming biological purposefulness of all behavioral attributes of creatures was so persuasive that Theodosius Dobzhansky, the Columbia University geneticist, in his book *Mankind Evolving*, even argued for the selective value of musicality, hinting darkly at the probable maladaptations in all the poor wretches who were tone deaf.

Things don't work quite that way. The human brain, probably from the time of the breakout onto the late Miocene plains anywhere from fifteen to ten million years ago, was an organ freed from closely honing selective conditions. Enclosed and pressed upon as it was by the anthropoid expansion of the Oligocene and the early Miocene, the hominid brain was originally a defensive organ.

Indeed when we go mad in our emotional response to threat or attack, defending children or country with fists and words, or in a thousand silent human passions and actions expressed in the heat of feeling, we are displaying an ancient mammalian-pongid intelligence. It is shrewd, quiet, stealthy, explosive, passionate, devout, and determined.

Once on the plains, man began to use this intelligence for aggression and opportunity. What was useful now was the cortical factor, the long-distance receptors, eyes and ears. Now intelligence had to "make it" in a whole new world of possibilities for which forty million years of forested primate existence had not prepared it.

Can we really say that "it just growed?" As we all know, change in the form and behavior of living things is made possible through constant gene mutation. Mutation rates vary. Living creatures that have found a secure niche—such as one of a host of one-celled animals floating in the sea, or mussels, snails, or clams settled comfortably at the bottom of oceans—tend to exhibit only very slow mutational changes. Various biochemical buffers to genetic change have allowed these creatures to remain virtually the same for many millions of years. Natural selection has operated during this time so that any change from their secure, inconspicuous role in the hierarchy of life has turned out to be detrimental. Since such deleterious change has caused the disappearance of variable types of clams, the category of "change" has disappeared from the clam family. What remain are clams that have very "stodgy" genes, i.e., they rarely mutate.

It is obvious that we humans have a unique place in evolutionary history. We are among the swingers who carved out their destiny by taking chances with the new. Here exists a much more dynamic rate of mutational change. The rate is only part of the story since natural selection will not allow pure anarchy in the kinds of mutations that can occur.

Embryonic development recapitulates in a brief and dramatic allegory the march of human evolution—we first live in a watery solution, briefly develop gills, a tail, and thence wend our way into existence as dependent little anthropoids. The reader will not be shocked to read my claim that it is doubtful that young apes will be born with scales or feathers instead of their typical down covering. Humans too, with some rare and unfortunate exceptions, will never be born with tails or webbed limbs.

**Directivity and Randomness in the Evolution of the Brain**

What has happened is that here too natural selection has eliminated from the repertoire of possibly positive mutational change certain combinations such as those mentioned above. By repeatedly striking down such combinations, it in effect has eliminated the individual carriers of certain mutation patterns. Conversely it has allowed to live and reproduce those individuals whose history of mutational change has worked for the good of that individual's ancestors, and thus his overall breeding pool.

George Gaylord Simpson has labeled this pattern of natural selection "orthoselection," selection in a straight line. This is where the brain comes in. For almost 150 million years, creatures with larger than average brains, which helped them to coordinate their long distance sense receptors on land, have been prospering. Pushing the line along sometimes slowly, sometimes more rapidly, this growing brain, along with an efficient structure of response mechanisms, has proved more or less constantly successful.

We speak as if mutations were random. Indeed they are. No creature can idly call forth some necessary or fortuitous mutation to help him survive. It is also not true to say that mutations take place in all directions. They do not. Mutation rates and the type of mutations that positively serve the destiny of a line of animals have tended to be selected out favorably. So while mutations occur "randomly," certain patterns tend to occur at a higher rate. Even mutation rates and directions are continually subject to natural selection.

With its history of service to the primates over that period of about forty million years since their origins in the great mammalian expansion, the unspecialized primate brain was obviously a prime focus for change in this class of animals. Naturally other factors made a wide variety of organs and structures useful. For man it was at first his incipient forest bipedalism that later allowed for a wholly new existence on the plains.

The restrictions on further development that existed in the forests because of the density of competitors and the lack of any strong selective applications for body and brain expansion, evaporated some ten to twelve million years ago. On his feet and roaming, this challenger entered a less constraining environment. The hominids simply had to wait for these mutations to show up. As the mutations began to appear, they were utilized in the adaptive process of surviving.

One last point about this process. Some organs or functions are simple, some complex. The organ that integrates the five senses—the brain—coordinates the extensive network of bodily functions and behaviors, and must be served by a consequently large network of genes. *Homo sapiens* has about thirty thousand genes which according to Sir John Eccles would involve $4.5 \times 10^7$ nucleotide pairs at the least. Many of these protein elements would individually control various brain and personality functions. Thus a mutation occurring at one point in a gene can alter a whole series of other features even though these are not under intense positive selection.

Like a rivulet of water exploring its way through the sands, the tentacles reach out to find the best route forward. Here too natural selection must favor variation so as to protect the experiment. If one approach fails, others will succeed. We have in this hominid move forward an explosion of brain expansion and restructuring as the pathways of adaptation go successfully and diversely forward.

There was thus also built into the high mutational rate an expectedly high degree of heterogeneity. Over time, going into the erectine stage, we became a variable animal form in brain structure. First we wandered along the highways of our African genesis, then into the intercontinental pathways of the world. No creature, no environment could now stop the evolutionary progress of mankind, no creature, that is, except man himself.

# IX

# The Children of Natural Selection

**An Ancient and Modern Truth**

It may be an old middle class parable, but it garners impressive evolutionary strength in its support: "we live for our children." For modern people, this is at least a twenty-one year commitment. Not merely that gift of life, sperm impregnating ovum. We must prepare our offspring for their own day as givers of life and security. The evolutionary dictum—the test of natural selection—lies in bringing the young of each generation to reproductive maturity.

Hold it, you say. Reproductive maturity begins at thirteen or fourteen years of age. True enough. Here already we have a smidgen of the mystery of *Homo*'s uniqueness. Merely being able to commit the act at age fourteen has never been a demonstration of one's maturity. At this age, a human is not experiencing a fully human act. "Making babies," as the street vernacular has it, represents a tragically degenerate condition in civilized life.

The nurturing of the human being, educating it, refining its behavior and skills so that it can enter the circle of humanity as a fully contributing member transcends the simple act of insemination by a light year. Literally it is the difference in millions of years of hominid evolution and behavior, for we have left the pongid stage behind. The children of natural selection, then, are the fully capable adults who are ready to defend themselves, their mates, and their children. The requirements are totally different from the "sex without responsibility" attitude, that tragic atavistic pathology that affects our contemporary culture.

This is not a religious sermon on morality. Yet isn't it interesting how we have transformed the evolutionary wisdom that lies below the threshold of consciousness into social rules and prohibitions? It is clearly something new in evolutionary history, the superego of restraint, discipline, slow building up of competency, unimaginable in that little blob of bigheaded, shrieking flesh that women labor so agonizingly to deliver.

## The Struggle Within

Let us attempt to discover how the natural, completely regular processes of evolution gave rise to such an unpredictable creature as man. In the process, we may find that the pongid, rat, wolf or ant models of adaptation and natural selection can finally be relegated to describing the evolution solely of these other honorable creatures. We humans have our model, a unique and powerful paradigm that has created an evolutionary channel as important as any of the great breakaway templates in all the history of life.

The brain has been a long time evolving. In serving its function down the taxonomic scale toward the fish, the echinoderms, lampreys, or worms, it was a humble, modest organ of survival. As it came into its own with the coming of the mammals, the brain became an increasingly important tool for natural selection. For a brief several-million-year period during the Paleocene sixty-five to sixty million years ago it was up a tree with the tree shrews and thence with the trunk-hopping tarsiers and lemurs.

The pace of brain growth and its selective importance continued to pick up speed from then on. The mutations followed relentlessly. To do so they had to have their feedback effect in ensuring the survival of their carriers. As we have seen, sometime during the Miocene, a small, furtive, defensive, intensely social creature, barely remaindered from the Oligocene ape-monkey parting of the ways, crept to the edge of the forest cover. It either held its breath and ambled tentatively forward or was booted out and told to run and get with it.

The mutations for brain size and structure were beyond any creature's ability to control and direct. Circumstance and chance in the name of natural selection were the sole and final arbiters. The real question is how this process shaped and directed this practically random process of variation to create first *Homo*, then *Homo sapiens sapiens*.

In order for this argument to go forward, we must understand a simple, clearcut declaration. Shortly after man ventured out upon the savannahs, he moved beyond the selective reach of other "competitor" animals. He may have been harmed on occasion by predators, even as are human beings today if they venture too close, for example, to a female bear with cubs or a prowling tiger in the jungle.

However, the key to survival in man was the defense of his "castle on the hill," his home, with pregnant mate and dependent children. It was the same then as it is today. The streets of our cities may be unsafe at any hour from the hominid predators that now stalk the shadows. We are uneasy and unhappy. The real outrage comes when our own homes are attacked or breached. When our society prevents us from defending them, those deep human yearnings of security are further outraged.

It was a moment reminiscent of the role of *Seymouria*, the primeval reptile that ventured out upon the raw land surfaces—a new and uninhabited world to conquer—finally free of the ancient aqueous dependencies. When man later trotted out into those open spaces, he breathed a brand new atmosphere of evolutionary opportunity. Those who had gone before had already accustomed themselves to their familiar diets and patterns of survival. He was alone.

Man's first competitors were his cousins, the well-formed land apes who seemingly also experimented with the open country life, the ramapithecines. His second were his own hominid brothers who were in joint exploration of this Garden of Eden. One does not have to reach for further animal metaphors on which to hang our evolutionary program. Man today is to a great extent as he was ten million years ago.

As he went forth and grew in accommodation with the new peripatetic life by long walks to fill an empty stomach, man saw where the competition was really coming from. If it wasn't a strict territoriality, the fact that he was of a wandering disposition made the transformation of brain energy into aggressive stuff easy. The relationships of the ferocious wandering Apaches on horseback against the peaceful Hopi farmers are still instructive.

It was not a cup of tea, vide the lives of recent Stone Age peoples, who have a great deal more cranial power than the Pliocene hominids of eight million years ago. Even then, however, *Homo* (in the making) had his basic repertoire of hominid adaptations: bipedality, sexual dimorphism, and powerful social-familial bonds, a human language communication system (probably also still in the making), and above all an intelligence that went far beyond that of any other line of mammal. These *Homo* had to learn to coordinate to survive. He had, in the mainline tradition of animal life, chosen the chancy, the open-ended path into an unknown future. In this he was also little different from contemporary humans.

## Freedom for Variability

During this long hegira through time, man had met none of the steady external biological pressures that created finely-honed adaptive organs—horns, hooves, fangs, extraordinary instinctual skills that were designed to fight off enemies or to gain power over a particular prey. Even today the peoples that have been pushed to the periphery, forced to live in climates and in ecologies that challenge the human body to its limits, have developed only secondary adaptations.

The Nilotic Negro living in an extremely hot climate has been selected toward a tall, thin body, a form conducive to shedding body heat. For the Eskimo, conversely, retention of body heat is crucial, thus his short, squat, plump body. The Incas of the altiplano live at twelve thousand feet and above. This was made possible by a unique adaptation, a mutant ability to pass more thinly oxygenated blood through the placental wall to the fetus. Women from lands at or about sea level would ordinarily abort a fetus under such conditions because the level of oxygen in their blood would be insufficient.

There are other minor adaptations. The pygmies are a major exception. Pygmies, living mostly in the tropics—the Laplanders are a possible exception—are examples of miniaturization, which seems to be a response to restricted ecological circumstances. In this regard, the status of the Laplanders is disputed. These groups reduce to the "size" of their environment, islands, restricted jungle environs, sometimes they reduce to meet the press of more vigorous and aggressive neighbors. Examples, among many worldwide, are the Congo pygmies as well as the paedomorphic Bushmen of the Kalahari Desert, about whom we will speak later.

The lack of external enemies or prey has kept humans generally unspecialized and has even given the human brain that extra bit of leeway to keep mutating in a wide variety of directions, but always broadly conditioned by the need for competency and social harmony. We can see this unity of mankind on the international scene where highly educated individuals have similar intellectual, cultural, and personal interests, even when they come from the four corners of the arth.

On the other hand, is there any doubt about the enormous variety of personality styles within a culture, or indeed the dizzying variety of cultures that are and have been? The tower of Babel was a Biblical symbolization of this fact. What in our biological nature has allowed for such variety? Further, why do cultures and their languages change in style and form over a period of time without real external reasons for these changes?

Innovation we call it. How does it make itself felt? The people of every new generation are biologically different human beings, and their differences stem in the main from their brains, endocrine systems, and other basic physical differences, however slight. Multiply the effects of one new human being even from a genotype that was once small—one or two dozen people and their billion sperm cells and ova—a million-or-so-fold and you get enormous possible variations that inevitably push the culture off in a new and always unpredictable direction. Nothing can ever stay the same with humans. It is the very force of their nature that will break the mold.

## Social Selection

An example may illustrate what is now the crucial point. In 1981, in a small farming town in Missouri, U.S.A., lived a bully. He committed many assaults, both verbal and physical, against individuals in the community. In several of these cases, he was let off on legal technicalities. Finally, after a siege of threats against the denizens of a town game room, he got his comeuppance. A crowd of men and women followed him and his wife out to his pickup truck. The people were enraged. They were convulsed with pent-up frustration against this man. As he started his truck, a shot rang out and he was killed. No one in this crowd of good citizens and town elders "saw" the person who shot the bully though the shooting had been literally in front of their eyes.

The bully's wife indentified the killer, but no one would corroborate her story.

In Greece, not too many years ago, a young matron in a small village, weak of flesh, succumbed to the blandishments of several casual male acquaintances. After it was learned that she was pregnant, possibly by the third of her "friends," several women relatives killed her. The incident was hushed up in the village and never came to the attention of the authorities.

These are extreme cases, but I would argue that for most of the history of our genus, a kind of natural selection took place that was not contradictory to the point underlined in the above illustrations. No question that much of the process of what might be called social selection occurred without the judicial deliberations of a council of elders, certainly without the principles of social order enunciated by Moses with his Ten Commandments.

Most often the selective sieve must simply have been one of general intellectual competence. Since we are more or less equal in general strength, sensory acuity, speed, and since most of the tasks we have to undertake in order to survive have to be done with others, the competency element, "holding up your part," counted for much.

Anthropologist Louis Leakey tried an experiment to investigate the problem of physical survival on the African savannah. He and several of his men caught and butchered a wildebeest. The smell of blood soon attracted a sizable crush of animals. Scavenger hyenas and wild dogs were almost beside themselves with desire. With only small stone knives, in imitation of hominids as lowly as the australopithecines, Leakey and company held off the animals by screaming, jumping, and waving their hands wildly as they hustled around the animal, awkwardly cutting off chunks of meat.

Such action does not require much social coordination, but certainly some bravery and efficient knowledge of how to cut and when to dance. Such behavior is beyond the ken of any predator prepared for the usual rigid animal defense and threat calls. This seemingly incontinent behavior—screams at all decibel levels—can confuse an animal, even when it is under enormous stimulus pressures, long enough for others to get the meat. Possibly, in the beginning, humans chased away as many competitors from already felled animals as they themselves killed.

Thus the process of gaining immortality for one's genes at all stages of hominid evolution had a significant social component. It was not achieved by the strongest individual grabbing his pick of the beauties and pulling her hair-first, Li'l Abner style, into the cave. Selection took place because of who married whom and whose children survived. Obviously, intelligence, will power, general capability, and aggressiveness led to leadership roles—and these qualities attracted the gals.

Women, as a rule, prefer acceptable men as sires for their children. Desirable women often have influential fathers and brothers. We need to emphasize the female element in the social selectionist equation. What is important is that *Homo*'s intelligence was broad-based; it mirrored his widening perceptions of experience. Humans love art, a good joke, sports, dance and music, admire technical skill, stealth, strength, verbal fluency. As human evolution progressed, all these characteristics must have figured larger as man became more secure. Just as *Homo* was a more able hunter and gatherer, he was likewise wiping from the slate of anthropoid evolution those creatures who in any way might have competed for the same or similar ecological niches as his own.

He did not have to commit genocide. Rather he could harry a tribe to the point that its reproductive capacities broke down. People on the run or in distress cannot calmly raise children. The Tasmanians are an example. In the mid-nineteenth century, the Tasmanian aborigines were herded onto a reservation on an island in the Bass Straits. By the

1870's the last pure-blooded Tasmanians had died off.

There are other selective hints from social history. All people have their individual colloquial names for the village simpleton—oaf, fool, jerk, schlemiel. This is the man you will not allow your daughter to marry. What kind of children would he give her? If the kids do not have brains or character, how will they make a living? My wager is that though *Homo erectus* may not have put it that way, circumstances made him act that way. When times are tough, as they were for about 9,999,500 years of man's 10,000,000-year tenure on this earth, the best gift you can give your children is intelligence.

If the most survivalistic act a person could commit was to insure that his mate's capabilities were at least level with his own, then certainly it was also in the social group's interest to foster a community of competent contributors. We celebrate the weddings of the best and the beautiful, for their good fortune will be ours. *Homo* would never have evolved if in his view the purpose of life in the social group was to take from it, to be supported, then give as a gift to it five slow-learning children who would in turn have to be supported. Such a process just could not occur without threatening the fabric of the society's survival and in consequence the survival of us all.

### The Usefulness of Intelligence

The variations for greater intelligence came episodically. Groups that experienced a high mortality rate because of severe conditions inevitably reflected a higher dynamic of change, thus a greater chance for high intelligence to survive and to spread more rapidly. This explains the discrepancy between the endocranial capacity of the northern erectines over that of the erectines existing in Africa and southeast Asia.

The Ice Ages were an extraordinary challenge to these

groups. The attrition rate had to be large. Because the groups were more isolated—conditions of cold would inevitably restrict mobility—inbreeding occurred more often, which added even greater probability to the concentration and emergence of only a few constellations of genes. If such an intelligence weighted the selective pointer just a few degrees on the plus side, it could result in group survival rather than extinction.

Over the long run, we know that when the chips are down the abler person will win. It won't necessarily result in dog-eat-dog behavior, as some of the genetic theorists of the "selfish gene" school would have it. Early on, self-sacrifice for the family, then for the group, had its selective value for close as well as distant members. Eventually, with the growth of the cortex—which was reflected in an increasingly complex society—altruistic cultural values could transcend even individual survival. Celibacy, fasting to the death, patriotic self-immolation are elements in human behavior that go far beyond any explicitly identifiable biological connection.

Humans live in a new world of values, values that clearly have their roots in the gradual transferrings within the crucible of survival and genetic immortality from the family to the larger social group. The crisis of modern life arises because the social net has been stretched to include large, imperial transethnic national societies. To the extent that these societies, living under the rules of abstract philosophical principles, forget the ancient truths of natural selection, they flirt with the same dangers of dissolution as did their progenitors.

Human intelligence is heterogeneous, for good evolutionary and adaptive reasons. Higher intelligence has been pulled out of the genetic sweepstakes to enhance the probabilities of individual, family, and ethnic-social group survival. The external morphological reality here is the consistent expansion of the human cranium and the concomitant improvement in tool design and fabrication.

Stone and bone are all that we have left to remind us of the social dramas of those forgotten eons. The hominid children of natural selection *were* increasingly brainier. Where did the others go?

# X

# The Erectine Finale

## An Evolutionary Margin

I believe that the evolutionary path that *Homo* took was inevitable. Why not? Intelligence, admittedly a complex and slowly-evolving adaptation, was one billion years coming truly into its own. Nature created titanotheres, the great squid, the blue whale, diplodocus, the boa, the condor, the mammoth, the tiger, all powerful, domineering creatures, each representing an extravagant adaptation. *Homo erectus*, big, bony, beetle-browed—his time came and he stayed the distance.

For about three million years, probably a good bit longer, our genus has been in the making. The classic erectine seems to have flourished in all of the great continental masses except the New World and Antarctica. Everything about the record of his tenure on the earth argues that his intelligence was more than adequate for all but the most devastating challenges by nature and nature's creatures. *H. erectus did* work. No one could have predicted what would happen afterward, literally the wrecking of the beautiful balance into which *erectus* had fitted. Why did this happen?

*Homo sapiens sapiens* is this other story. *Homo* now had a soul. The theologians are correct in making that distinction between man and nature. I would draw the line

between erectines and "supersapiens." It was with that las
bit of consciousness—that extra margin of awareness tha
created in *Homo sapiens sapiens* an awareness of mortality
his mythological reactions to life and death, his ability t
discern the existence of evil, to be able to choose betwee
alternatives, to consider consequences—that *Homo sapien
sapiens* acquired a soul. Here religions and the religiou
became weighted with depth and humility.

How did we get from then to now? Evolutionar
biologists argue that successful species usually block thei
own pathways to evolutionary progress. Successful i
utilizing their adaptation, such species spread out, becom
panmictic (random mating), perhaps differentiating sligh
ly into semi-adaptive races. Yet they remain interfertile
Having carved out and dominated a particular ecology, the
are loath to let it go. That domination often proves to b
their downfall, for they specialize or get too set in their way
With time comes change; then there is no one to bail then
out—rather like a giant overgrown transcontinental societ
like Persia, Rome, or even the United States. Eventuall
there was no place to go for a breath of new evolutionary ai
It is a story that has repeated itself over and over again for a
least a billion years.

*Homo erectus* having its repertoire of hominid adap
tations, technology, language abilities, cultural coordina
tion, and strength should have stayed on for at least five t
ten million years (the average tenure for any species), a goo
solid endurance record for mainline creatures. Instead, *H
erectus* quietly disappeared with the coming of sapient mar

*H. erectus* did not just grow smoothly into *Homo sapien
sapiens*. Note the fragility of *Homo sapiens sapiens*: bor
bigheaded, completely helpless for so long, slow to matur
hairless, delicate of bone—a skull covering of bone that i
many individuals is less than 4 mm. thick.

Compare this to the evolution of modern horses. Whe
one knows of the fossil evidence of their smaller, earlie

ersions, the later forms are not surprising. The same goes or elephants, apes, and cats. To expand the five-foot erectine o a six-foot version and to lose all that bone at the same time ust doesn't seem probable. Something radical had to have happened. In what must have been an evolutionary blink, Mother Nature produced one of the most untraditional creatures ever ladled from her bubbling evolutionary brew.

## Languid Evolutionary Rhythms

Consider the time factors. By about 250,000 B.P. the almost imperceptible stabilization of the erectines was progressing throughout the Old World. Peking man's successes admittedly had resulted in a bigger-brained form, but without any radical alterations. In Africa and southeast Asia, the pace was even more languid. Perhaps nature was saying *mañana* to the urgings of intelligence and the occasional popping up of something intellectually exceptional: "Bananas and mangoes we have. So, pick one, stay in the shade. Tomorrow we'll try to figure out what to do with you, smart one!"

However, in Europe we do find skeletons that hint of something askew. The Steinheim lady found in Germany, of ca. 200,000 B.P., had a respectable endocranial capacity of about 1200 cm$^3$; at the same time it was a hominid, somewhat more delicate than the usual erectine. However, this was probably a female. (Female skulls tend to be both smaller and more delicately-boned than male.)

Somewhere within the same time frame came the Swanscombe skull cap, found near London, England. Still probably erectine, although possibly a Neanderthal ancestor, this fossil shows indications of having an endocranial capacity in the range of 1300 cm$^3$. Associated with the Swanscombe cap were typical Acheulean tools. The brain was definitely enlarged at this advanced moment in history, 200,000-250,000 B.P. Yet this human creature was still compatible with his ancestors.

The mysterious and seemingly unsteady rate of advance is brought home in a now puzzling bit of fossil bone found at Vértesszöllös in Hungary. This posterior skull section shows the typical configuration of erectine morphology. It is older (400,000 B.P.) than the previous skull fragments and its ruggedness perhaps reflects this age. Yet the endocranial capacity indicated by the skull bones points to a range of 1400-1450 cm³, possibly even to 1600 cm³, which would put it well over the sapient border.

It should be emphasized that we are still working with pieces of fossil bone, possibly deformed by the events of many millennia, subject to the mental reconstruction of contemporary scientists. Often estimations of the total endocranial capacity vary. Indeed, we can only assume that among all the guesses, with so many variables, a mean consensus can be arrived at.

Recall the events at Torralba in northern Spain at about 300,000-400,000 B.P. A tremendous kill took place presumably by *Homo* erectines. We can't be sure, because no human bones were found at the site. All that we know is based on the evidence of charred animal bones. Seemingly campfires were used to cook the food and/or ignite fire brands. The area around Torralba was a swamp at that period, and the supposition is that erectine hominids had stampeded elephants and deer into the swamps where the animals wallowed helplessly in the thick gluey mud. There they were killed and an orgy of feasting likely took place.

It seems probable that the swamp became a great erectine encampment, for the piles of animal bones indicate that much food was taken and regularly consumed over a long period of time. Even at that early stage in human development, certain perennial human characteristics are hinted at—indulgence in feast and then famine. The creatures seem not to have stored the meat away. Perhaps they did not know how to, or would not. Another quality is he seeming orgies of killing. If the great northern mammals were not yet exterminated by *Homo erectus* it was not because of lack of exuberance.

The erectines were cannibalistic when necessary and given greater capability would have been massive destroyers. This is what is fascinating about these shadowy progenitors of sapient man. Wandering out of Africa about one-and-a-half to one million years ago, they maintained the quest for sustenance and survival for well over a million years. These were humans. They were limited, they were different, but they probably showed no less intelligence than the lowest levels of competency in our world. In fact the northern erectines had larger endocranial capacities, despite the primitive structure of their skull, than some recent peoples of simple cultural existence.

During this long period, the erectines experienced not merely the warm swampy interstadial periods, like that original one that perhaps drew them out of Africa for the richer, untrammeled pickings in the northern latitude. By contrast, the classic erectine of Java and the Indies, *Pithecanthropus erectus*, lived in a true Garden of Eden. In one sense this was a tragic evolutionary circumstance for the latter, for it placed him forever in a backwater existence from which his sapient successors have still not removed themselves.

In the north the ice came and receded several times during these hundreds of thousands of years. Buffeted and wandering, *Homo erectus* endured. His capacities were more severely tested; the natural heterogeneity in brain size spun out its slow advance, keeping him one step ahead of the evolutionary scythe.

Wherever we find erectines, one fact seems to come through. His numbers were never large. On the three continents there were perhaps no more than one or two hundred thousand, enough in this proportion to nature for him to have existed with and in nature for a very long time.

Having destroyed his pongid and hominid competition, *Homo erectus* was now a racially variable, yet interbreeding superspecies, not extraordinarily numerous, yet completely in control of the selective conditions, given any external event except the most cataclysmic.

## Last Stop on the Main Line

*H. erectus* is important because he tells us about the limits of our exploitations of nature. He lived in the bosom of nature. Not shut up on islands or enclosed in the dark jungle valleys, *Homo erectus* wandered widely through ancient Europe, Asia, and Africa.

*Homo erectus* was a creature of culture. His social capacities had not yet overwhelmed nature. He lived in the rhythm of the external flow of the seasons. Much of what we are as hominid mammals, in our emotions, sense of unity with nature, love of beauty was probably solidified in the inchoate sensibilities of this creature. Because his brain had not yet given him the momentum to break out of his cocoon, he lived within his cocoon, and for an inordinate period of time.

If ever we are forced back to nature, to its harshness and unpredictabilities, we can use *Homo erectus'* experience, for he survived the dangers of the north and the wild fluxes of its seasons. Don't use the natives of the jungles or the island paradises as models of the natural. They are the atavists of the modern period; they are endowed with the spark of sapiency yet are for the moment prohibited from joining its northern dynamics. These peoples are probably caricatures of what life in the raw grip of nature was or might be. The difference is that even then, 250,000 years ago, *Homo erectus* was straining at the bit, roaming, struggling, surviving. Perhaps he was experiencing in these last, anxious, pregnant centuries the thumping birth pangs of his successors' impatience to take up the challenge.

The erectine tradition could have gone on for much longer, but it was not destined to. New events and a new kind of man were incubating. The indelible imprint of *Homo erectus*, however, is still within us. In fact, he *is* us. We could probably interbreed with *Homo erectus*, at the least with his final incarnations. Just as we have bred back to resurrect the Shetland pony, we could resurrect *Homo erectus.*

Here and there we have seen him, not pure, but intermixed, a complex of features, perhaps occasionally inhabiting the body of a very intelligent individual. Here and there, we see his chin, his brows, an odd-shaped skull, indeed even the halting intelligence and simple words. If we put them all together, we could see ourselves 500,000 B.P.

Of course, we cannot really go back now, because we have arrived; we have achieved those irreversible results that *Homo erectus* strove for, though impotently. Thus if consciously or through stupidity we undo the work of evolution and return to a state of dependency and biological embeddedness, we will find a wholly new reality—a desert.

The animals, the flora, even the soil, are almost gone now. The irony is, we really *can't* go home again. *Homo erectus* was the last stop on the trunk line of primate evolution. The gorillas, orangutans today experience this reality in their pathetic existence. Going back to nature would probably herald our extinction.

We climbed up beyond *Homo erectus* to a higher cliff. We started to explore it. A dark recess opened up to receive us. When we finally emerged 200,000 years later we were at the edge. In back of us was only a sheer drop. Since then we have been busily eliminating any possibilities for other egress. We can only go forward now. To hope for success in the perilous climb, we must rediscover those primeval laws that brought us this far.

# XI

# Into the Tunnel

## Explosive Sidetrack

It is almost like a magician's trick. Into the hat goes the handkerchief. A moment later, out pops the rabbit. Or else picture it as a tunnel. The dauntless old engine, huffing and puffing with steam and coal smoke, goes into one end of the tunnel and comes out the other end the sleek diesel 20th Century Limited.

That is the picture of human evolution that many scientists now have to face. That "gradual" transformation, the delight of stodgy traditionalists, a montage of motion picture frames, fulfills the view of evolution as proceeding by small incremental mutations. Now selecting in a fairly straight line, nature transforms the pumpkin into the glass coach.

Humans are one of the numerous exceptions to this smooth process. Otherwise, as I have noted above, we would still be tough, bony, chinless little erectines. Something happened within the tunnel, a period of perhaps 150,000 to 200,000 years, a period of epochal impact on the fate of life on our planet.

In this chapter and those that follow I will outline a train of thought and evidence that if not necessarily unorthodox at least has not been spelled out in the particular configuration that I am going to give it. Remember, no part of evolutionary theory is *not* subject to controversy. (See Bibliography, Chapter XI.) The key to all the conjecture and hypothesis is "will it fly?". Will it "bake bread" as a set of ideas that explains other facts or the new evidence as it comes in? I think that what follows will prove to be fairly close to the truth and, what is more, go a long way toward better explaining our contemporary biosocial circumstances.

There have been many great leaps from regularity in evolution: the vertebrates, amphibians, birds, mammals and finally man. I say "man," or rather "*Homo sapiens sapiens*," and not hominids because I think that an intelligent mammal was inevitable. That it proved to be a primate and not a fissiped (whale) is as you know also a matter of great controversy. That is, whales may be equally intelligent and capable in their own world as were the australopithecines or even the early hominids. Both lines coexisted in their respective worlds without turning things upside down. Not so *Homo sapiens sapiens*. About that the reader has about as much evidence as I.

Just catalogue the physical differences that this new creature had, a creature whose actual bones we find first from a period as late as 35,000 B.P. Immediately we see his cultural impact. *Homo sapiens sapiens* did not wait his proper one million years to begin to show his stuff. *Homo sapiens sapiens* may very well have been fully formed earlier, but for reasons yet unknown, his presence is only indirectly indicated. Sometime within that tunnel of two hundred thousand years he was transfigured.

We see emerging from it a variety of human effects, almost preludes or appetizers anticipating the main course when man would finally arrive. From that moment on, the story is his and in ever more sweeping and concentric circles of impact. Is it still going on? You bet your life it is. Thirty-five thousand years of activated *Homo sapiens sapiens*, but only about ten thousand years of life beyond the Ice Ages.

The story has barely begun.

## The Challenge of the Ice

Contrast our new super *sapiens* with *Homo erectus.* *Homo sapiens sapiens* is almost one foot taller. He has a chin. Rather than keeling sharply to the sides from its center ridge, his cranium rises to its greatest height over the ears. In the back it falls abruptly to be tucked in behind the spinal cord. His forehead is freed of those massive ridges that so constricted the frontal areas of the brain of *Homo erectus.* Instead *Homo sapiens sapiens* has a high brow, his forehead rising sharply over the eyes, thence soaring high over the parietals like a balloon. No erectine depression behind the frontal bony ridges here.

*Homo sapiens sapiens* has relatively little body hair. The teeth are extraordinarily delicate, the bone structure extremely fragile, even when the creature has a height of six feet. The cranial cavity is enclosed by a bare four to five millimeters of bone, no longer the thick skull of *Homo erectus.* In short, here is a big baby. Everything points to the fact that an enormous remodeling job has taken place during the 150,000 to 200,000-year blackout.

The turn toward the mammalian format from the original therapsid dog-like reptile some two hundred million years earlier had a similar dynamic and structural character to this deviation of *Homo sapiens sapiens* from the erectine prototype. I would bet that the earliest hominid Oligocene ancestor, also at a time of instability, was a somewhat bizarre departure from most of the evolving pongids. Given time, these forms moved away from the orthodox modeling into a world of their own. Subsequently they came back to deliver a fatal blow to their progenitors. Little first steps of difference sometimes led to enormous gaps, induced by a secondary level of competition. First, time was needed, then a cause, and finally the force were needed to carve out this separation.

What could have occurred to cause this infant giant to come into being? One requirement was separation, the sometimes desperately lonely path that creation requires. The genes had to stay where they were, not be dissipated within the general mass of erectine life. Second, a blow was necessary, an occasion for a powerful speeding up of selective pressures. Now every possible variation became a desperate roll of the selective dice.

Probably these populations diminished radically. Every variant became a crucible of change, epochal in itself because even if it affected only a small family, this root social unit could become a meteor of integrity protecting the future. This is not Lone Ranger Lamarckism, where need stimulates the "just right" instant gene coming to the rescue. In evolution, "need" produces nothing. Fortunately, variation is not completely random and thus the eon-long reservoir of struggle and success had eventually to reassert itself.

Remember, back in the late Oligocene and early Miocene, the ancestral line of the hominids probably took a paedomorphic step into obscurity to escape the anthropoid avalanche, viz the foreshortened muzzle, the small teeth, the powerful family bonds, the energetic magnetism of sexual love and dimorphic devotion, the quiet murmurings of whispered conversation.

Could some sudden chill from the oncoming Riss glaciation (ca. 200,000 B.P.) have trapped several unfortunate erectine wanderers in the north and forced them to deal with a totally new climate, perhaps even an unfamiliar geography? There is no question that great change arises from extreme testing, tribulations that ultimately involve survival itself. To that ancient crossopterygian lung fish: either escape this drying pool and hope for another one close beyond the bank or take a deep breath of that stuff up there and try to stick it out. Whatever necessitated their decision—

choice or circumstance—humans must have suffered in toughing it out.

## Baby Face

They say that paedomorphosis, the retention of childlike physical and behavioral characteristics of the ancestral form into the adult stage of the descendant, can present radical adaptive opportunities for a line of animals. It can break up the usual selective flow of events. If creatures retain infant characteristics into adulthood, these unusual physical and behavioral features are novel to the outside world. External pressures now affect the paedomorphs differently than they did their unpaedomorphized predecessors. Perhaps the paedomorphs are more playful, more experimental, even more physically resilient.

Another possibility, given the appearance of such a radical genetic aberration, is that certain features continue to grow slowly well into a creature's maturity (caenogenesis). Perhaps, as in the case of man, as evidenced in the more delicate bone structure, the period of growth was lengthened in time. Could it be that the bone structure of incipient sapients, the heavy brow ridges, the rugged jaw did not develop into full-blown adult form? If so, could this be the clue to the brain's balloon-like growth, unimpeded by a bony blockade, as in the erectines?

The infant chimpanzee is much more human in appearance and physical structure than the adult. The foreshortened face, without the prognathous muzzle, is also infant-like. Even the closing of the sutures of the skull, which is completed in the chimpanzee by about six months, in *Homo sapiens sapiens* is delayed until adolescence. Could the extra time for brain growth in *Homo sapiens sapiens* have been the source of that extra measure of adaptive potential that made survival a possibility?

What we observe is a switch from a comfy adaptive

stance, polished and secured by over a million years of practice, to a tumultuous, perhaps even frantic grasping for the right selective combination for survival.

Let us reemphasize that paedomorphism often leads to covert evolution, a creative opportunity, because the shift in development is hidden from the traditional selective pressures facing the species. For example, the critical moment of survival in every species is at the time of maturity, sexuality, and mating. If the mature members have difficulty raising their young to reproductive maturity, the future becomes insecure.

If, on the other hand, the sexually mature adults now act like kids, they present a different face to the natural enemy. Perhaps they elude them completely because they are a different color or are smaller, e.g., dwarfs. Maybe they fly at a different time of day or scavenge on the forest floor rather than higher up. What is important is that they can now reproduce and thus survive. Examples of failed paedomorphosis can be found in the platypus (monotremes) and even in dwarfed humans, especially the highly paedomorphized Bushmen of the Kalahari Desert. While it doesn't always work when carried to its ultimate structural and behavioral conclusions, "rate" gene variation is part of the repertoire of mutational changes carried down the ages in our genes.

Paedomorphosis as a selective option probably appeared under conditions of crisis in certain populations of incipient *Homo sapiens sapiens*. Evidence reveals it to have been a specialized manifestation of this process. Instead of following the path toward smaller size, *Homo* remained tall and got even taller. The selective dynamics of the north were different. Larger territories needed to be covered for food. (This would have been more difficult for small people.) Could it also be that the great northern mammals could best be hunted by big strong people? Is it possible that to pass a child that would have a large head, the female pelvis had to enlarge further? Anthropologist Irvine DeVore has argued that the transition to bipedalism had as a by-product the widening of the female pelvis. Certainly a systematic set of restructurings united here to create new opportunities.

Whatever actual physical changes occurred in man, one selective factor stands out: the 1500-1600+ cm$^3$ brain capacity of the Cro-Magnon Euro-Asiatics when they finally stepped from the tunnel in about 35,000 B.P. It is that highly restructured, paedomorphic cranium for which all the other synchronic remodeling of the human body was carried out.

The brain allowed man to fend off the challenges of animate and inanimate nature. It freed him from the fixed cultural patterns of the erectines and allowed him to ask those silent and not so silent questions with which children are so persistent: "Why, why, why?" For each question that man put to nature, he had to propose possible answers. It could be a matter of life or death. This child mind had to be able to recognize the failure of an action to fulfill the "need question," whether it was where to find animal food, or where best to locate a shelter when moving the family group to a new location. They say young minds are far less rigid. Just as young mammals play to learn—sometimes posing novel solutions that might have eluded the fixed and unimaginative response patterns of their elders—so too with our paedomorphic founding fathers. *Homo sapiens sapiens* was now capable of asking a series of hypothetical questions about life and survival and to keep trying out new possibilities.

Man could do this and learn in a way that erectines could only dimly imagine. The neurons were there in excess billions. A longer and more learning life could in turn be passed on to the young. Indeed a whole sequence of feedback possibilities would inevitably have its consequences for survival.

**Conclusion**

In summary we see in *Homo erectus* an established creature, yet one who held within himself the possibilities, even the realities, of morphological change. Some erectines entering the frigid incubator of the north sometime after 500,000 B.P. were caught in a cyclone of selective pressures. The challenges, most likely climatic, geographic, and ecological—a new kind of weather, new lands, new natural balances in food and shelter—probably decimated those populations to some extent. Each rapid variation in this line of mammals, which already had a highly dynamic mutation rate, was significant. It was imbued with a powerful adaptive punch. Most likely more succumbed than survived.

At first those key paedomorphic characteristics were probably maladaptive. Likely they did not constitute enough of a negative imbalance to do these transitionals in completely. The restructuring thus led to a small covert population of evolving "erectine sapients." Eventually, either a climatic easing or else the gradual working out of these novel characteristics released the newly created sapients from their northern cocoon and their genes flowed out, rushing with adaptive confidence, a trickle that was to become a torrent.

# XII

# Tipping the Scales

## Controversial Lady

Gleaming white, it lay amid the random detritus of the mine excavations. At first, by the workers, a reaction of revulsion and fear. Some terrible deed unearthed, an investigation, trouble. Then, as mine supervisors as well as workers gathered around, it became clear that this skull was different. It was human, but.... Ever alert to the possibility of discovering fossil bones and aware that a contemporary burial ground would hardly be found that isolated and deep in the earth, the discoverers were soon apprised of the truth.

This was in 1925, near Broken Hill in what is now Zambia. The skull was probably a female, hence first labeled Rhodesian lady. It was obviously a late hominid. Beyond that, controversy raged about its age. The skull had been discovered only after the site had been excavated and thus its *exact* place in the underground strata was in question. The best guess at that point was that it was quite recent, possibly fifty to twenty-five thousand years old. Its mineralization was complete and the tools that were subsequently found in that area—fairly rudimentary flints and choppers—conformed to the general run-of-the-mill Wilton technology of that part of Africa for that late-to-end of the Pleistocene period.

The real puzzle was the skull itself. In retrospect its significance stands out even more clearly in relief. The endocranial capacity was fairly large, about 1280 cm³. The brow ridges, however, were the largest of any sapient yet uncovered, an enormous bony Mount Everest towering over the cranial carriage behind and the face in front. Little did this lady from Zambia know that she would set off a controversy that could unhinge one of the cornerstones of hominid evolutionary theory.

This would not be another Piltdown hoax with its endless conjectures and arguments. The Piltdown skull—human cranium, ape jaw—had been disturbing. It was there and had to be accepted, even if it did not fit theory or other fossil "facts." The actual repute of the Rhodesian skull has never come into question. In fact a whole series of associated fossils from that general geography of southern Africa, as well as that broad stratigraphy, has since come to light. The problem lies in the age of the fossil and its relationship to what else was going on in human evolution during the late Pleistocene.

The burning question was whether Rhodesian man was yet a sapient. Carleton Coon, one of the most distinguished, also one of the most controversial, paleontologists, firmly argued that it was well within the erectine tradition and in fact far behind in morphological characteristics some of the more advanced specimens in both Europe and China. (However, these were at least several hundred thousand years older.)

That was it. At the time that the Rhodesian lady supposedly lived in southern Africa, fully hominized supersapients were already or soon would be painting their magnificent murals on the cave walls in such sites as Altamira in Spain, Lascaux in France, and other points from the Atlantic to the Ural Mountains deep inside Russia. Such a picture of discontinuities in human evolution would deeply disturb the traditional assumption of the uniform hominid flow upward in time at all points of the geographic compass.

What added even more haze to the clarity of perspective was Coon's famous analysis and conclusion in *Origin of Races* that these African fossils were ancestral to the Negro, just as the European erectines were ancestral Caucasoids, and finally the specialized northeastern Asiatic erectines (Peking man) were ancient Mongoloids.

## Genesis and Change

The African equation in the understanding of human evolution is vital. Human evolution seems to have begun in Africa; certainly it was here that it took its first great steps forward. It was only after *Homo erectus* picked himself up sometime over one million years ago and went wandering that Africa became a backwater of human evolutionary progress. The reason for this is not hard to understand. It may help to consider that, despite what the Europeans may want to think, it was the fresh invigoration of progress that took place in the newly discovered lands in the sixteenth to eighteenth centuries A.D. that enabled Europe to maintain its strength and power.

Similarly, it was the progressive surges of energy and innovation that arose from places such as North and South America, Australia, New Zealand, and Canada that kept that circle of invigoration going throughout the modern period. Indeed, as we will discuss below, the exact same relationships occurred when the progressive northern tier spiraled its influence into the south (Asia as well as Africa) that brought the south once more into the mainstream of evolutionary events.

An example of this relaxed pace of evolutionary progress in Africa can be seen in the comparison of Oldowan IX (or Chellean III), an early erectine skull of about one million years ago. When we compare it to the Rhodesian (Broken

Hill) circle of skulls, it is clear that there has been some expansion of the cranium. It is also clear, as David Pilbeam notes, that the basic structure has remained strikingly stable. An explanation lies in the advances and retreats of the ice packs in the north. Africa was somnolent, though some alterations had to have resulted in its basic climate. When the ice advanced into southern France and central Italy, Northern Africa was veritably a garden—with much rainfall, a rich carpet of green, and many kinds of animal life. Could it have been that way in Roman times? During the interstadials (when the ice withdrew), if the evidence of recent events is any guide, northern Africa suffered the dry winds from the East.

However, southern and central Africa north to the Ethiopian highlands still had the warmth, often the rain, and the untouched Pleistocene ecology to fall back upon. Life was probably easy for man, made especially easy with the adequate erectine brain structure. The pressures being off here in Africa, natural selection was more languid about who was allowed to survive while, in contrast, it was pushing for radical progress in the north.

Change does come. Its portent is intriguing both for what it says in itself and for what it implies about events elsewhere. A series of skulls has since been unearthed from various geographic levels throughout Africa from the period 125,000-75,000 B.P. In fact, it is because of the new evidence that the respected opposition has gathered its forces to reclassify the Rhodesian fossil as belonging within this latter time frame rather than the much more recent one formerly proposed.

Dating fossils is one of the most controversial topics imaginable; it is always "subject to emendation." The various methods—carbon 14, potassium argon, uranium decay—are still approximations; it is difficult to make a crucial test because the material is so distant in age and the margin for error is large. The best one can do is to devise a kind of locus of dating probabilities that arise from the different methods, then to factor in, along with the paleogeological evidence, the rock strata, and finally the logic of the bones.

It is not beyond possibility that in his enthusiasm a scientist may exorcize his own judicious doubts about where the material was found, probable associated strata, and dating approximation by the above chemical methods, by allowing the conclusions to fit into whatever prevailing view of the evolutionary process he seems to favor at the moment. Thus we can understand the battles that go on among the various contestants. Not only a good idea may be at stake, but the fame and fortune that go with being the winner of the debate, at least for one year.

Let us line up the African skeletal evidence ca. 1984. It seems to divide into two groups: lagging erectine-sapient types (Ndutu, Bodo, Broken Hill, Saldanha) and borderline or early sapients definitely on the way over the mark (Laetoli 18, Omo I, Omo II). (See Bibliography, Chapter XII.) The consensus is that about 100,000 B.P., more modern populations were arising from the ancient African erectine model, but now with a rounder cranial configuration, larger brain size, and even some reduction in the bony mass typical of our erectine forefathers.

Here and there a hesitant conjecture is put forth that, suddenly, new genetic elements must have entered the scene, because within this approximately fifty-thousand-year slice of evolutionary history there is too much differentiation, too many variations on the old theme. Also, no other supporting evidence would argue for significant ecological or climatic disturbances. The large populations (otherwise we would not have so many fossil examples) seem to argue for stability within change.

The key phrase concerning the late-Pleistocene African scene is "genetic infusion from without." Where did these genes come from that were gradually herding the local erectine stock into the sapient corral? The guess is they came from the north, where the progressive erectines had long outpaced their African compatriots in the march toward a sapient morphological Rubicon.

## Contact at a Distance

For another suggestive tidbit let us go halfway around the world to the East Indies. Here too, man had been slowed up in the selective expansion of the cranial container. As mentioned earlier, the pithecanthropine erectines in southeast Asia were consistently behind their northern Chinese contemporaries, often by as much as 200 cm$^3$ in endocranial capacity, but toward the end of the Ice Ages, the Pleistocene, quickening of the pace of change can also be noted.

These southeast Asian erectines had evolved as a related subspecies of man, now classified broadly within the Australid racial taxon. One day in 1930, on the island of Java, a fossil skull came to light. The area had been rich in fossil materials: one more would not raise eyebrows. This one, however, eventually dubbed "Wadjak man," was an exception. Typical of the evolving Australid erectines, it was heavy-boned and archaic, but its brain case was extremely large, indicating an endocranial capacity of 1400 cm$^3$. The dating places it anywhere from 40,000 B.P. to 100,000 B.P. Could it have been one of those fortuitous anticipators that takes hold selectively and pulls the entire group forward? Else, might Wadjak have been the herald of a new source of genetic progress from the outside? We cannot know for sure.

Yet man is and was a wanderer. To traverse the enormous distances, man used nothing more technically advanced than his feet and his long-perfected bipedalism. We moderns ought not be so smug about our capacity to cover distances. To us time is in short supply. Nature, however, proceeds at its own deliberate pace. The Pleistocene was at least 1.75 to two million years long. Plenty of time to walk many miles, with children, pregnant wife, and perhaps even a few old ones.

The Papuans, on the island of New Guinea, are examples. They are settled primarily in the western highlands. In their racial origin, the Papuans are Australids. Indications from both fossil and recent evidence argue for genetic continuity in that part of the world. The people who live there now are in all likelihood descendants of those who lived there many thousands of years ago, with perhaps one little difference: A few strange sets of genes may have entered from time to time.

The Papuans are atypical Melanesians: Their facial configuration is relatively aquiline, with large convex noses—very different from the flat, spread, nasal physiognomy of the Melanesian area Guineans. Normal variations, we might say, except it is a striking exception to the rule in what is a basic physical trait of these southeast Asian natives. The Papuans also on occasion exhibit a pinkish skin, more often reddish hair. Certainly no recently wandering Oriental could have been responsible either for the aberrant proboscis or the color variations.

A variation having little selective value that suddenly appeared and took hold in a small population would not be an uncommon evolutionary event. Therefore the possibility exists that the genes came down at generations removed—as in the game "telephone"—but watered down. Yet the nose perdured and lodged itself with the unsuspecting Papuans. There was all the time in the world for this to have happened.

### Inevitable Miscegenation

Now to my hypothesis about the process of becoming human. Let us start at the end, which we know now was only the beginning. The Cro-Magnons (*Homo sapiens sapiens*) arrived in the European caves in about 35,000 B.P. They were a glorious and paradoxical people. In every area of life, which we will describe in a moment, they were extraordinarily brainy, passionate, exuberant, creative. They were full of heaven and hell. At 35,000 B.P., however, they had not just been created. For a good period of time before that, they must have been in the making.

Go back in time another forty thousand years to 75,000 B.P. Here, all over northern Europe and western Asia, ranged the Neanderthals. Sapiens, true, but not *Homo sapiens sapiens*. They were a more primitive people. As we will describe them in the next chapter, they were close in appearance to what one might expect were we to raise *Homo erectus* by about five or six inches, expand his cranium, and do a modest amount of reorganization.

Of course the Neanderthals didn't just arrive in 75,000 B.P. either, because like the Cro-Magnons they didn't change much from that time on nor did their culture. They were a stable, if varied race. A reasonable conjecture would be that they were on the scene, if not in great numbers, for a period before their burst of affluence and expansion.

That could bring us back close to 100,000 B.P. Since the further back we go the more iffy the time factor's accuracy, we can say that much was happening all over the world from about 125,000 B.P. to the present.

My guess is that here and there, as various northern erectine stocks took the full brunt of the Riss glaciation (from 250,000 to 150,000 B.P.), the effects of this crisis radiated. Perhaps a few half-formed sapiens made their way south and did what all humans have done, spread their genes as they went, like pebbles rippling the water. As we well know, the direction of gene flow cannot be predicted. It would be natural for a small band of people with chattering teeth and frozen toes to head as far south as possible, given the consent of the resident hominids.

I remember once looking at my half-Jewish infant

daughter, then at my all-Jewish father, noting the Oriental tinge in the eyelids covering my daughter's blue eyes. I glanced at my father's eyes and his high cheekbones. From where and how far back did that Russian intrude into our close family circle? Farther back still, how many generations removed did the wandering Mongols sweep into his domain?

Inside the tunnel of travail and suffering far north of those Ethiopian highlands at Omo, epochal events were occurring. Even at 100,000 B.P. the great movement must have been in progress, the scale of intelligence and hominization moving irrevocably off its then secure center

# XIII

# The Neanderthal Experiment

## First Sapients?

We no longer look upon them as brutes. They are our brothers, yet it is highly doubtful that if we could, as Carleton Coon once suggested, get them to wash up and put on suits, we would see them melt into the morning rush hour in Zurich. Not even a haircut or shave would aid their quest for anonymity. The title *Homo sapiens neanderthalensis* hints at the problem they present to us about the implications of their sapiency.

Some scientists believe that the range of human variation that they represent falls well within what exists today. In fact their overall endocranial capacity is quite a bit larger than that of contemporary sapiens on the lower slope of the worldwide curve of human brain size. They are perhaps one of Mother Nature's last experiments on the sapient model, but one that she was allowed to follow through until the final intervention of man.

The process still continues, but it is no longer a matter of spontaneous creation and destruction, as it was with the Neanderthals. Rather it is a question of conscious human intentions. Indeed, everything depends upon how the human community deals with the enormous variety of human types, cultures, and interests.

The Neanderthals arose and disappeared, were slaughtered and absorbed. The message of this for us takes us one step closer to the modern dilemma of sapiency. Let us take a quick look at these people, what they were, and how they provide us with another glimpse into the enigmatic evolution of human nature.

## The Mystery of Neanderthal

We have known of the Neanderthal fossils since the time of Darwin. Discovered throughout the length and breadth first of Europe and then western Asia, they were the first "ape-men." Today we know that we interbred with them. If you have a long, thick, aquiline nose, a la Charles DeGaulle, perhaps way back, several Neanderthals were in your family tree. In short, if you are a Caucasoid, you probably have their genes in you.

At this point in the controversy, this question comes up: were they ancestral to *Homo sapiens sapiens* or were they collateral sapiens, another and parallel line of creatures that emerged from the Mindel-Riss interglacial period, after ca. 150,000 B.P., but having undergone far less remodeling than the classic Cro-Magnon type? One could hedge one's bets by saying that we derive from one of the more progressive Neanderthal types, those that inhabited eastern Europe and western Asia.

It was the German and French Neanderthal fossils that really gave them their bad name. As the earliest examples to be uncovered, the west European Neanderthals provided much stimulation for the artistic imagination. Out came a slumped-over, 5'4", prognathous brute, heavy-boned, almost arthritic-appearing in its awkwardness, a caricature of what the Neanderthals perhaps really looked like. On the other hand, as one examines the various Neanderthal skulls, east and west, and compares them with the "typical" erectine type, one may come to a striking conclusion. Blow up the

| ERA | SYSTEM AND PERIOD | SERIES AND EPOCH | DISTINCTIVE FEATURES | YEARS BEFORE PRESENT |
|---|---|---|---|---|
| CENOZOIC | QUATERNARY | RECENT | Modern man | 11 THOUSAND |
| CENOZOIC | QUATERNARY | PLEISTOCENE | Early man; northern glaciation | .5 TO 3 MILLION |
| CENOZOIC | TERTIARY | PLIOCENE | Large carnivores | 13 ± 1 MILLION |
| CENOZOIC | TERTIARY | MIOCENE | First abundant grazing mammals | 25 ± 1 MILLION |
| CENOZOIC | TERTIARY | OLIGOCENE | Large running mammals | 36 ± 2 MILLION |
| CENOZOIC | TERTIARY | EOCENE | Many modern types of mammals | 58 ± 2 MILLION |
| CENOZOIC | TERTIARY | PALEOCENE | First placental mammals | 63 ± 2 MILLION |
| MESOZOIC | CRETACEOUS | | First flowering plants; climax of dinosaurs and ammonites, followed by extinction | |
| MESOZOIC | JURASSIC | | First birds, first mammals; dinosaurs and ammonites abundant | 135 ± 5 MILLION |
| MESOZOIC | TRIASSIC | | First dinosaurs. Abundant cycads and conifers | 180 ± 5 MILLION |
| MESOZOIC | | | | 230 ± 10 MILLION |
| PALEOZOIC | PERMIAN | | Extinction of many kinds of marine animals, including trilobites. Glaciation at low latitudes | 280 ± 10 MILLION |
| PALEOZOIC | CARBON-IFEROUS | PENNSYLVANIAN | Great coal forests, conifers. First reptiles | 310 ± 10 MILLION |
| PALEOZOIC | CARBON-IFEROUS | MISSISSIPPIAN | Sharks and amphibians abundant. Large and numerous scale trees and seed ferns | 345 ± 10 MILLION |
| PALEOZOIC | DEVONIAN | | First amphibians and ammonites; fishes abundant | 405 ± 10 MILLION |
| PALEOZOIC | SILURIAN | | First terrestrial plants and animals | 425 ± 10 MILLION |
| PALEOZOIC | ORDOVICIAN | | First fishes; invertebrates dominant | 500 ± 10 MILLION |
| PALEOZOIC | CAMBRIAN | | First abundant record of marine life; trilobites dominant, followed by extinction of about two-thirds of trilobite families | 600 ± 50 MILLION |
| | PRE-CAMBRIAN | | Fossils extremely rare, consisting of primitive aquatic plants. Evidence of glaciation. Oldest dated algae, over 2,600 million years. | |

**Diagram 1.** The Geological Ages.

The geological ages are dated through the analysis of remaining radioactive elements at different levels. There is a significant margin of error.

(From Norman D. Newell.)

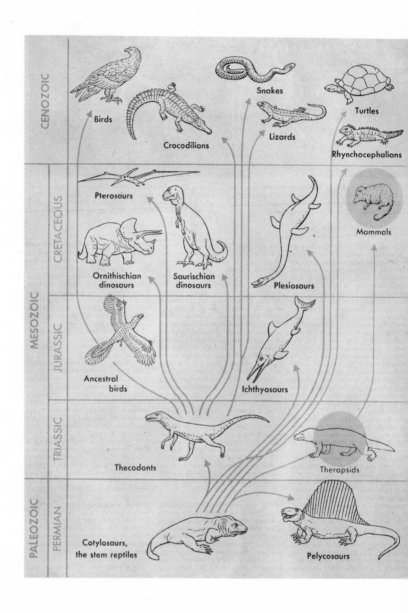

**Diagram 2.** Origin of the Mammals.

Relatively unsuccessful at first, the mammals made an end run around the explosive dinosaurian advance. At the end of the Cretaceous, the mammals probably gave the dinosaurs their comeuppance.

(From E. H. Colbert.)

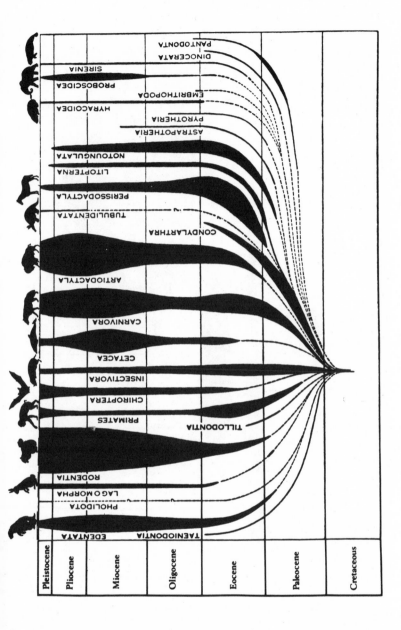

**Diagram 3.** Evolution of the Placental Mammals.
The primates up until the recent period represent only a modest proportion
of mammalian life. The dominance of any one line is often followed by its
decline.

(After Romer.)

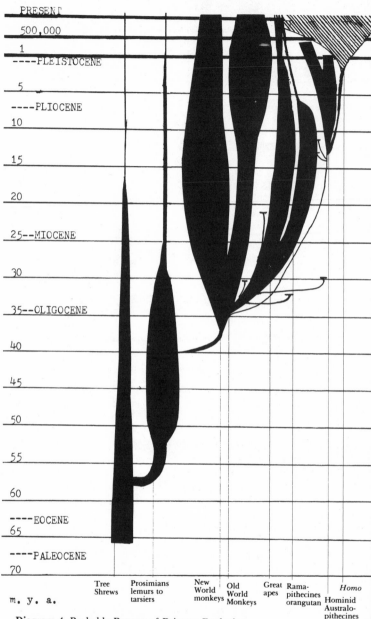

PRESENT
500,000
1
----PLEISTOCENE
5
----PLIOCENE
10
15
20
25--MIOCENE
30
35--OLIGOCENE
40
45
50
55
60
----EOCENE
65
----PALEOCENE
70

m. y. a.

| Tree Shrews | Prosimians lemurs to tarsiers | New World monkeys | Old World Monkeys | Great apes | Rama-pithecines orangutan | Homo Hominid Australo-pithecines |
|---|---|---|---|---|---|---|

**Diagram 4.** Probable Pattern of Primate Evolution.
Note a: progressive decline of the shrews and prosimians, once more advanced anthropoids evolved; b: the dominance of the apes into the Miocene is short-circuited by the evolution of the hominids. A side product of ape decline was the expansion of the Old World monkeys; c: the hominids have their origins as one of a number of anthropoid dissidents from the main line; d: *Homo* probably separated from the australopithecines between thirteen and ten million years ago.

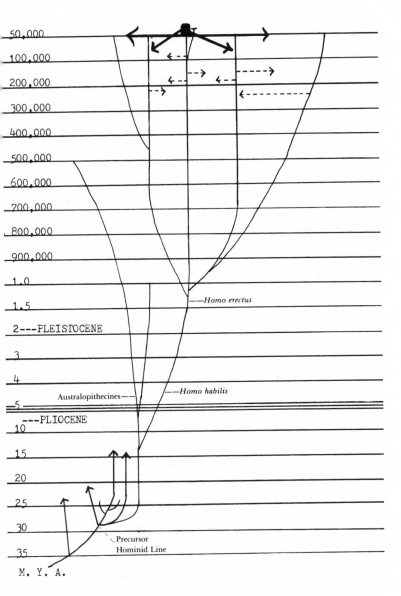

**Diagram 5.** Probable Pattern of Hominid Evolution.
Evidence seems to indicate the migration and separation of the erectines from 1.5 to one million years ago. It is suggested that continuous, if small scale, genetic contact was maintained by the erectine subspecies throughout their tenure. *Homo sapiens sapiens* was extruded from European-West Asian Caucasoid advanced erectine stock sometime after 250,000 B.P. By 100,000 B.P., *Homo sapiens sapiens* genes were tilting mankind over the sapiens line and creating the modern races of humankind.

**Plate I.** Venus, Willendorf on the Danube, Austria.
Aurignacian, about 20,000 B.P. Made of dolomite limestone. 11 cm. height.
Some of the original covering of red ochre can still be discerned on this
artistic masterpiece. See Chapter XVI.

(Natural History Museum, Vienna)

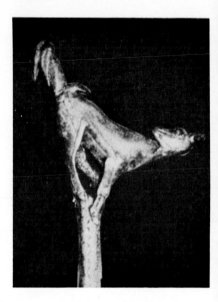

**Plate II.** Magdalenian Spearthrowers.
These spearthrowers were made of reindeer antler during the last and
bitterest part of the Würm glaciation, 12,000-10,000 B.P. These "practical"
tools were decorated in the manner of medieval warriors. The example with
the fawn defecating is especially noteworthy because of its scatalogical
humor and brilliant esthetic and technical realization. The other is an
evocation of a ptarmigan. See Chapter XVI.

(Grahame Clark and Geoffrey Bibby)

**Plate III.** Rock Wall Paintings.
The similarity of these vivacious realizations in brilliant reds and other colors points to a Capoid presence. At top, a hunting scene from Castellan in eastern Spain, end of the Paleolithic. Bottom, late nineteenth century, shows Bushmen rustlers defending their take from Bantu pursuers. See Chapter XIV.

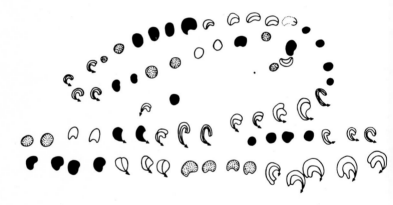

**Plate IV.** Inscribed Notation.
Example of a tradition of markings found throughout the Cro-Magnon peoples' range. These markings on a palm-sized piece of bone date to the mid-Aurignacian, southwestern France, ca. 30,000 B.P. Marshack believes that they represent a two-month span, tracing the waxing and waning of the moon. See Chapter XVII.

(Alexander Marshack)

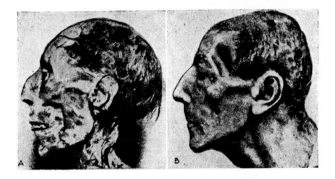

**Plate V.** Mesolithic Migrations.

As the ice receded, peoples were on the move, yet cultural traditions remained. Top: From Denmark to Ofnet, Germany (bed of skulls with attached neck bones) to earliest Jericho, 8000 B.C. (plaster-encased skull), old methods of ceremonial burials perdured. Bottom: remarkable similarity of a: mummy of Egyptian Pharaoh Ramses the Great (1225 B.C.) and b: death mask of Prussian Frederick the Great (1786 A.D.). See Chapters XVIII, XIX.

(Grahame Clark, Franz Weidenreich)

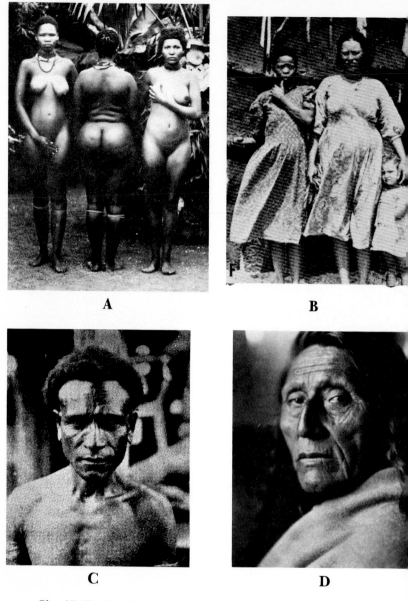

**Plate VI.** The Hybridization of *Homo*: an Ancient and Ongoing Process.
1. a: Hottentot women. Traditional fatty deposits in buttocks can be observed. Woman on right is an offspring of a Caucasoid father, Hottentot mother. (Musée de l'Homme, Paris)

b: Three generations. Bushman mother, her half-Caucasoid daughter, the latter's three-quarter Caucasoid daughter. The youngest has seemingly lost all traces of her Bushman ancestry. (P. V. Tobias)

c: New Guinea Papuan with a non-Melanesian nose. (Musée de l'Homme, Paris)

d: North American Plains Indians. The Mongoloid-Caucasoid heritage is patently evident here. (Brace and Montagu)

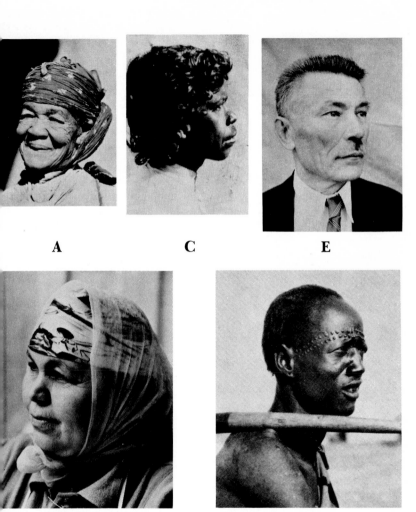

A        C        E

B        D

**Plate VII.** Hybridization 2.

a-b: Morocco. A woman from southern Morocco showing Hottentot (Capoid)-like features and a broad-faced Cro-Magnon-like woman of the Rifian hill country.

c: Southern Arabian boy (Hadhramaut) with Australid features.

d: Shilluk (Nilotid Negro, southern Sudan, Tanzania) with Caucasoid elements.

e: Hungarian with reminders of an eastern Mongoloid heritage.

(All from C. Coon, *The Living Races of Man*)

**Plate VIII.** Finale, 1859.
Group picture (Oyster Cove) of the last Tasmanians (Australid-Melanesian heritage). The last survivor, Truganini, center row, right, died in 1876, aged 70.

(American Museum of Natural History)

erectine skull, lengthen it front to back, greatly emphasize the occipital bun (a bulge at the back of the skull), smooth out the keeled indentation of the skull coming off the center ridge, do the same behind the heavy brow ridges, and one has, essentially, a Neanderthal.

It is a big skull, with room for a big brain inside. Still, as the old Lucky Strike ad used to put it: "It's what's up front that counts." The Neanderthals were still low-brows. Their culture, as indicated by their artifacts, bears this out. (Note, for example, by comparison, the Acheulean's—*Homo erectus*—beautiful hand axes.) The Neanderthals were adept at slicing flint from cores to make a variety of blades and scrapers. Some of their encampments are virtually awash with flints. Yet throughout the approximately fifty thousand years that their skeletal remains cover, hardly any cultural development is indicated. Over and over again anthropologists have wondered at the technological mind set of these peoples, their lack of experimentation and innovation. Compared to what occurred immediately after, even as compared with some of the Acheulean remains, the Neanderthals, big brains and all, did not progress. Their culture was simply inert.

If we put the two dates side by side, we find a revealing sequence. The earliest Neanderthal fossils are dated at about 75,000 B.P. Now remember, the African borderline sapiens date from 125,000 to about 75,000 B.P. Practically all physical anthropologists agree that the Euro-Asian Neanderthal model is quite different from these African skulls.

The Cro-Magnons, the most highly sapientized hominid, arrived on the scene in France about 35,000 B.P. Oddly, the first skeletal material seems to be associated with Mousterian or Neanderthal stone tools. Could these *Homo sapiens sapiens* have merely taken over the tools or patterns previously used by the Neanderthals whom they had just driven from the caves?

Great differences exist between the classic (western) Neanderthal fossils and *Homo sapiens sapiens*. Even postcranial (below the neck) structure exhibits a thickness and massiveness unlike *Homo sapiens sapiens*. My guess would be that these were two associated races that made themselves at home in this northern Eur-Asian tier. One would represent an incomplete remodeling of the ancient erectine prototype, the other a literal revolution in physical structure and brain type. That at later stages these two groups may have met is likely. The result would be, as always, extermination and/or miscegenation. The east European Neanderthaloids may possibly have been saved from the extremes of the Neanderthal model both because of their contact with a wider variety of hominids to the southwest and because of a somewhat less drastic adaptation for what was necessitated in the cold of western Europe.

A possible example of hybridization between Neanderthaloid and *Homo sapiens sapiens* types is given in the enigmatic Tabun and Skuhl skeletal remains in Israel. The former date from about 60,000 B.P. and seem to be mainly modified Neanderthal types. Their tools are also Mousterian. However, in their more rounded skulls and slightly more gracile bone structure they seem to point to an extragenetic element.

Near Haifa are the more recent skeletons of Skuhl. They date from about 40,000 B.P. Among these variable Neanderthaloids is Skuhl V, an odd-shaped skull, which is not only different from the others found in the same cave, but one which does not fit into either the sapient or the Neanderthal camp. Most anthropologists classify Skuhl V as a hybrid. Is it possible that this unassociated creature had a Neanderthaloid mother and a Cro-Magnon father? The mystery is why we find no Cro-Magnon bones or culture until that sudden and seemingly irrevocable takeover of the northern Eur-Asian caves. Did climate finally force the Cro-Magnons to take cover after they had wandered for thousands of years searching for security, scattering their genes here and there with the passing generations?

## An Erectine-Sapiens Cross?

Let us consider another scenario. It may emphasize how tenuous our knowledge is and how many different explanations at this point in our ignorance are possible, indeed have to be entertained. We certainly find in the Neanderthals a real break in the cultural continuity of the late Pleistocene. Kenneth Oakley has claimed to be able to trace parallel encampments and lithic (stone tool) industries of the Acheulean and Mousterian cultures. The peoples were literally passing each other by. However, the margin of encounter might have been as much as several thousand years. This seems to argue that, wherever the Neanderthals were created, they moved into a western Europe already occupied by the old-time Acheulean-advanced erectines.

We do not find any European erectine skeletal material this late, 100,000 to 75,000 B.P. If some had survived the climatic horrors of the first Würm peak 120,000 to 110,000 B.P., these erectines were probably soon exterminated by the bigger, tougher Neanderthalers, who were excellent hunters. Parenthetically, that would be enough of an explanation to clarify the Neanderthalers' lack of technological progress. They had a successful thing going for them in their hunting outfits and skills. Also, times were too tough to experiment to go beyond what worked. Again, just a hypothesis.

Another guess about the origin of the Neanderthalers is that they represent a *Homo sapiens-Homo erectus* cross at the middle stages of the *Homo sapiens sapiens* crucible of genetic and selective transfiguration (200,000 to 150,000 B.P.). Further, this cross possibly became solidified over several thousands of generations through intensified inbreeding to form a stable type quite distinguishable from the southern (African) erectines and the holed-up, incipient

Cro-magnon *Homo sapiens sapiens.*

Remember, at the level where the fossils of Neanderthal appear, no sign of a European erectine has been found. Nature's first complete sapient experiment was on the way, even while the human world of man was being far more slowly transformed below the Tropic of Cancer. We ought not to think of this first sapient as a brute. Already signs clearly point to what this enlarged brain of man would create from the indifference of nature's processes.

## A Soul in the Making

The caves of western Asia, at Shanidar, (Iraq), 46,000 B.P., and Teshik-Tash, 40,000 to 35,000 B.P., tell us about the fascinating world of this human. At Teshik-Tash, seventy-eight miles south of Samarkand, almost in the Biblical world of man's creation, the Neanderthals may have lived under momentarily benign circumstances. A child had been born and shortly died. A grave had been dug at the cave entrance. Around the grave several pairs of antelope horns were symmetrically placed. Atop the grave a profusion of pollen grains was found. From a distance of thousands of years, the guess is that the grieving Neanderthals placed flowers on the beloved infant's grave.

At Shanidar, a few thousand years earlier than the above, a cave-in had occurred. A young mature male adult was killed, and lay buried under the debris. Forty-five thousand years later, the cave was opened; the bones had become fossilized. The individual had an atrophied leg. This had apparently been caused by a catastrophe that had occurred when the man was young, for this leg had stopped growing long before the young man's adolescence. That he had survived the first accident and even though obviously of little use to the band, being lame, he was kept alive and succored until the tragedy of his adulthood, testify further to the band's sense of devotion and philanthropy.

A soul was flickering in the Neanderthal sapiens, a dim light of empathy and caring. Now, they were freed of the automatic mammalian responses of instinct to protect, then abandon. A sense of a future, of planning, of commitment beyond survival needs gained importance. Indeed, the winds of tomorrow blew over those frosty northern Persian plateaus. Mankind was in the making.

## The Neanderthal in the Modern World

The Neanderthals no longer exist as an entity. Probably they were driven from their cave refuges by the oncoming Cro-Magnons, a fate similar to that suffered by earlier hominid forms. Partly it was slaughter, partly they were harried into reproductive impotency or were absorbed to some degree through miscegenation. The harsh climatic conditions did not abate from 35,000 B.P. In fact, toward the end of the Würm glaciation they began a final frigid crescendo.

What remained of the Neanderthal genetic heritage in the body of northern Caucasoids was probably partially eliminated through assortative mating. That is, those members of more lethargic or limited mentality were gradually excluded from the family-forming circle. Indeed the Cro-Magnons may always have had cultural prohibitions against absorbing captured Neanderthal families into the group, or even allowing their mixed-blood progeny to live.

Nevertheless, one has to assume that the Eurasian hunting ground of both the Neanderthals and the Cro-Magnons was big and empty enough to allow for many mixed-blood peoples to coexist; gradually they would be absorbed into the basic European Caucasoid type. We know that intelligence is highly selective and half-breeds were probably more able than the relative percentage of parental blood would indicate. At a place called Kara Kamar near

Haibak in northern Afghanistan is a cave without bones, just tools. The top level shows evidence of Mesolithic habitation ca. 10,000 B.P. Below that are the remnants of a flake culture, typical of that created by the Neanderthals. The date for this is about 35,000 B.P. during a momentary warm period of the Würm glaciation. Just below this flake level are blade tools, evidence that Cro-Magnons may have *preceded* the Neanderthals, by how many passing centuries?

That a Neanderthal element is within us and that the Caucasoids to this day remain a highly variable genotype physically, intellectually, and culturally lend some credence to the above conjectures. Just as we can breed up, we can breed down. Here again, our heritage can become our future.

Could the Neanderthals have survived in our modern world? For tens of millions of years, endocranial capacity has been the key element in assaying the intellectual competency of a species to deal with a varying environment. The Neanderthals had an endocranial capacity of up to about 1500 cm$^3$ and sometimes larger. However, according to Philip Lieberman, they had a different vocal structure from ours, in that their ability to produce vowel sounds was restricted. At first Lieberman argued that this restriction in articulation would have hampered their language abilities and cut down on learning and communication. Björn Kurtin has postulated that the apparent early deaths of adults (females at twenty-five, males at forty years of age) probably hindered learning by youngsters since their parents were either old, senile, or dead as the young reached puberty. Therefore, despite their large brains, they could not progress culturally.

Most languages use between twenty and fifty distinct sounds for communication. A language can use as few as two phonemes (dot-dash of Morse code), or as many as one hundred. Were the Neanderthals to articulate even a dozen distinct sounds, these could have been incorporated into a highly specific system of human communication. Also, were we to have brought the Neanderthals into our world, with modern medicine, it is likely that we could have extended their life expectancy so that their young could be responsive to a modicum of educational stimulation.

Finally, many of the primitive peoples found in the nineteenth century had skulls with far less endocranial capacity than the Neanderthals, though with a more typical sapient structure. Their skulls, however, may sometimes have vestiges of brow ridges and thicker crania. For instance, the Australian aborigines, the extinct Tasmanian aborigines, and the latters' related New Guinea-type Melanesian peoples often have an archaic cranial configuration with endocranial measurements that sometimes dip below 1000 cm$^3$.

How far above these kinds of gross physical parameters, given the archaic Neanderthal skull pattern, would a hominid need to function independently in our highly abstract technological world? What could these peoples have learned to do to become economically independent? After all, some humans living in the modern world are carried along on the wave of affluence and protection that civilized societies provide.

That the Neanderthals did not advance culturally in their more than fifty thousand years that we have been able to trace does not tell us how they would use what others, more modern than they, would have provided for them. At any rate, it is moot now. We are a highly mixed variable species, only a short distance down the evolutionary road beyond our Neanderthal brothers. The total distance that we have come is long, the distance beyond the horizon unknown.

# PART 3

## Arrival

# XIV

## Man of the Future?

**Brief History**

Loren Eiseley called him "man of the future." This was Boskop man, whose fossilized remains were one of a series found in the southwest Transvaal of South Africa in 1913. The Boskopoids, as they are now more generally called, are truly remarkable specimens. They seem to have come out of nowhere, out of context of any slow, incremental evolutionary stages. Boskop man just suddenly appeared in the late Pleistocene (30,000 to 20,000 B.P.?), with highly infantilized skull features, a tiny face, overpowered by a huge swelling skull, and an endocranial capacity of often well over 1600 cm³, in some cases over 1700 cm³.

Eiseley asked how did he get there? What did his sudden southern African presence mean and where did he go from there? It is a mystery, for though the Boskopoids left descendants—the Bushmen and the Hottentots—these tragic peoples are a far cry from those truly portentous fossils. Seemingly a great future, finally a lamentable denouement.

Little consensus exists thus far on the peripatetic fate of this taxon of humanity. Beyond an apparent agreement on its common Capoid racial affinity with present Bushman populations, as noted above, no positive conclusion has been arrived at to indicate that this group is a truly independent racial strain. For one, the racial integrity of the Capoids is too recent as compared with the other races (only to the end of the Pleistocene, not long before we encounter Boskop).

Some authorities even see them as a Negroid variant, perhaps like the Congo Pygmies, who also seem to depart from the Negroid pattern, and who, like the Bushmen, have become dwarfed, though without the paedomorphism of the latter. Let us go back and trace the wanderings of these people both for what they seem to tell us about the enigmas as well as the known nature of man.

## Wanderings of a People

For the beginning we have to go north once more to the frigid incubator where the final sculpturing of the human form occurred. In this case, north is only southern Spain. We find here the artistic signature of these people, a clear indication that they were and still are sensitive and creative sapients. On exterior rock walls (unlike Cro-Magnon, who preferred the dark interiors of caves), these Capoids drew their lovely evocations of their experiences. In one case it is a sketchy, two-dimensional drawing of a young girl, basket in hand, obviously in search of honey. In another it is a vivid encounter with an unseen enemy, the protagonists dynamically wielding bows and arrows, whether in attack or defense, it it not clear.

These drawings—colorful, sketchy, and highly active— are quite different from the powerful cave paintings of the Cro-Magnons. Form and action are realized in a special esthetic shorthand, which epitomizes the Capoid style from this point on. The evidence from the drawings is that the Capoids were in conflict with someone else. If so, with whom were these internecine battles?

More probably, the conflicts were with the resident Cro-Magnon bands. In one Cro-Magnon painting is a grisly representation of a cage basket hung from the edge of a cliff. In the basket is a human being, imprisoned like a trapped animal. The characteristics of this person are clearly depicted and he is obviously a Capoid. The cruel humor of this painting and the events that it describes imply a kind of racial disdain, which if correctly interpreted shows how long and deeply established were those human cultural and psychological barriers between "we" and "they."

Soon the evidence for Capoid habitation in Spain ends. The story is picked up in northern Africa in the widely dispersed Aterian culture. Though the technology of these Capoids was advanced, it was equal neither in esthetics nor design to that of the northerners. Fossil evidence for the correlation, however, confirms the association. Farther south, in the now desert areas of the Sahara, are additional indications of these people. A number of caves, probably either encampments or long-term residences, display a continuing vivacity in the rock paintings—for example at Tassili-n-Ajjer, south of Benghazi—discovered by Henri Lhote in 1957.

Suddenly, the Aterians were displaced. The so-called Moullians arrived. The time frame is still obscure, but it was probably between 30,000 and 20,000 B.P. Who were the newcomers? They seemed to be those broadfaced Cro-Magnons from the north who moved from east to west along the coast of the African continent. One should not be surprised to find indications of a migration of Moullians south through the Strait of Gibraltar—perhaps along the same passageway as the Capoids on their experimental invasion of Europe.

The genetic imprint of both populations still remains. Here and there, especially in the southern border desert areas, the memory of the Capoids is evoked—wizened bodies, high cheekbones, wrinkled, sallow skin. Hidden as recessives, these genetic elements still coalesce and come to the surface, a glancing memory of the past.

We see traces of the Moullians more often—broad faces, blond hair, blue eyes. They were the newcomers. (Perhaps the Berbers, living in the isolated recesses of the Atlas Mountains, with a separate language and ethnic identity from their Arab lowland neighbors, carry these traces.) Soon, however, even the Moullians stepped back into unwritten history.

A newcomer, most likely from the east, a delicate variant on the northern Caucasoid model, one that would dominate the early history of the Mediterranean, overran the culture of the broad-faced people. We would recognize these newcomers. They were slender, olive-skinned, had long narrow faces and black hair. They have inhabited the area since about 15,000 B.P.

What about the Aterian Capoids? Where did they go? Carleton Coon argues forcefully that they were driven south into the open areas of Africa, then still inhabited by remnant and transitional hominid types. Two pathways south were open, one directly south through the then watered savannahs of the now dry Sahara, the other over the east African highlands via the Nile River. The latter has long been the route for northern invaders, and much later the way back for slave traders of the ancient and medieval world.

The so-called Singha man of 25,000 to 20,000 B.P. in the Sudan was a Capoid. This individual was large-brained and tall, but with an incredibly thick (13 mm.) bony covering to his skull for so recent a human. Other seeming Capoid fossils have been found, which also may indicate that this highland route was a Capoid pathway, but little cultural or genetic evidence remains of their passage. Thus it may have been an uneventful, even unnoticed *Völkerwanderung*.

The only evidence of the other alternative—a pathway directly south through the Sahara—is in their rock paintings. Of course these settings could have been only seasonal encampments for hunting or exploring during a time when the Sahara was likely a real Garden of Eden, with water, many animals and birds, much fruitful flora.

How many years did these Capoids wander? The Israelites wandered for forty years, which seems a long time to us. However, we must remember that the major part of human history is written in man's migratory instinct to pick up and search for better circumstances. Could this instinct have been passed on from generation to generation to keep a people constantly walking until the chief, like some primeval Brigham Young scanning a new and tantalizing valley, would call out, "Here it is, here we stop"?

They may have wandered two hundred years, perhaps a thousand. That's a drop in the bucket compared with the spans of time with which we measure the evolution of man. However, at last, deep in southern Africa, again living on semi-temperate highlands, are found the Boskopoids, apparently well settled in. What happened here we will never know. Whether in the process of their contact with Moullians along the Mediterranean shoreline they acquired new genetic elements that helped swell their brain case, we cannot know.

## Gentility and Decline

The Capoids today show a distinct Mongoloid cast in their physiognomy: Oriental eyes, high cheekbones, and sallow, tannish skin. Coon conjectures that possibly a *Homo erectus pekinensis* contact may have occurred in the dim past. However, no tangible evidence has been found to indicate that such an erectine form ever migrated from its northeast Asiatic homeland until very late—thus in the modern sapient stage of evolution.

Perhaps some unknow crisis, climatic, ecological, even a final human challenge, may have stimulated the evolution of a human being with a delicate physiognomy that was unparalleled even in the north. Coon had some doubts, saying that the Boskopoids were well within the late-

Pleistocene sapient range of variation. Eiseley wonders whether they came upon the earth too soon. He hints that their culture may have mirrored the seeming gentility and peaceability of the Kalahari Bushmen (today we know that even the Bushmen can be provoked to violence) and that rougher, less developed hominids did them in.

Could such large-brained creatures have been that helpless? Consider the Cro-Magnons. There is little evidence of cannibalism among these later peoples as compared with the erectines and even Neanderthal man, but everything else indicates that the Cro-Magnons were a tough people. Even the pattern of cultural replacement that is evident in the caves argues for a wandering people capable of expelling the indigenous dwellers from their winter quarters. This could not have occurred peacefully.

The Boskopoids never seemed to have developed the technology that might have been expected from their futuristic cranial conformation. On the other hand, neither did the Mongoloid sapients, who were from roughly the same late-Pleistocene period and who also had already made the transition to an advanced sapient morphology. Could some remnant morphological throwback have lain hidden beneath the cranium that prevented the Boskopoids from using their intelligence to defend themselves more effectively? Could their advanced cranial structure have given us a false impression of their complete sapientization? Finally could the hidden defect have been a paedomorphism that did not work, that threw the creature back into a physically youthful structural state, now possibly to block the kind of intellectual and behavioral maturity that might have allowed him to develop a more outgoing and creative culture?

The art of the Bushmen even today is strikingly reminiscent of that which decorated the southern Spanish rock walls of twenty-five thousand to thirty thousand years ago. Doubtless the connection between the noble diminutive, harried Bushman and his ancient Ice Age predecessor is established. Today, the Hottentots are already partially interbred with Negroids and ancient Caucasoid-Negroid cattle traders from the north. However, the racial distinctions can still be seen: the extraordinary buttocks of the female, the peppercorn hair, the labia minora genital extension of the female (enormously · lengthened and adhering). All these physical characteristics are different from both their Negroid as well as their Caucasoid neighbors. Such characteristics in their soft parts would of course not be evident in the fossilized skeletal material.

An ancient race, whose history is shrouded in the multiple enigmas of distant eons, seems about to disappear. Eiseley called Boskop "man of the future." Eiseley felt that the gentleness, the delicacy, the intellectuality preordained in this fascinating people would someday make itself universally felt. I think that Eiseley meant that we ought not assume the tragic inevitability. Intelligence has been thrusting its way into the future for too many millennia to be thus daunted. No, the Bushmen and the Hottentots may represent a marginal line of descendants, but the Boskopoid heritage is out there waiting for us to perceive and identify once more.

**Man of the Future**

I sat in my office. Before me, animatedly rocking in my Salem chair, was my friend Frederick K. (real name changed). He was visiting from South Africa. A black professor in a black university, he had been allowed only grudgingly to visit the United States. We were speaking off the record. At this moment, in his typically enthusiastic, guileless way, he was discoursing on his people, the northern South African Sotho who had migrated many generations ago from the north, pressing both the Bushmen and the Hottentots before them.

When the whites came into South Africa between the seventeenth and nineteenth centuries, they found the Hottentots barely hugging the coastline, already interbred with the oncoming blacks. It was obviously a vigorous migration of peoples that met in South Africa, both black and white. The few Capoids gradually slipped from the world's consciousness.

"The Zulu to the south of our people are warriors," said Frederick. "They enjoy the hoopla of military rhetoric," he laughed good-humoredly, but with some condescension. "We Sothos are more interested in education." He spun off a long list of elders of his tribe as well as his noble contemporaries: neurosurgeons, attorneys, scholars, writers. "Our ordinary people are running South Africa. We are the electricians, the auto mechanics. South Africa works because of us. The white sits in his office supervising," he chuckled. "When they are not dozing, they are by their swimming pools enjoying the good life. We Sothos work and study."

I looked at him admiringly. A short, chubby man full of good will and toleration. His hooded eyes glistened, the high cheekbones were full of action, layering a prematurely wizened face. A man well into his mid-forties, he still had a "baby-face" cast to his physiognomy. "My God, Frederick," I muttered to myself, "damned if you're not a Boskopoid!"

# XV

# The Middle Kingdom Moves Outward

Would you be shocked if I said that the twenty-first century will be the era of the northeast Mongoloids—Chinese, Japanese, Koreans? This claim is far more persuasive today than it would have been a scant generation ago. A long historical record of civilization, a highly advanced sapient morphology—big heads and brains, an already observable and outstanding functional intelligence (high I.Q.)—are the key elements. That these groups have innumerable social and economic problems ought not be too daunting. In human evolution the northeast Mongoloids are a surprise and somewhat of an enigma. Let us probe the mystery that inheres in their not too portentous sapient beginnings. Their experience will serve as a factual example of how we ought to keep looking forward at man rather than back at what is really a series of historically irrelevant racial separations in human evolution.

## A Private History

The consensus today is that the Mongoloids, of all the races, seem to have had the most distinctive, even private, history. Of all the races, they are the most different. The destiny of the Negroids and Caucasoids has long been intertwined. The routes into Africa and Eurasia have been well trodden. Capoids have struggled with both Caucasoids and Negroids throughout their history. At the very beginning of our awareness of sapient dynamics, the Capoids and Caucasoids in Europe and northern Africa made contact. Then at the very end, into our time, the white southern Africans pushed the Capoids around and isolated them. The Negroids coming south almost into the modern era have overrun the Capoid homelands, perhaps at an early stage sharing some genetic contact.

Even the Australids, though a distinct race throughout their history, had confused us in their racial affiliations. We had believed that the Polynesian and Maori, for example, even the Hawaiians, were Mongoloid-Caucasoid hybrids. The Ainu of Japan had long been thought to be an ancient and recidivistic Caucasoid people.

Now these stocks are seen as basically Australid crosses, harking back to the remnant "pure" Australid peoples of New Guinea and the islands (Melanesians) as well as the Australid aborigines and their cousins, the extinct Tasmanians.

Some see in the general configuration of the Australid body and facial structure a rough similarity to the Caucasoids. Perhaps the genes that suddenly sent the late-Pleistocene Australids over the sapient threshold arrived from the Caucasoid homeland in western Asia. We know that Caucasoids were living in India (the Tamil languages are evidence) and along the Indus River long before the Indo-European Hindus came down from the north between 2000 and 1500 B.C.

Since the Australids are a varied people, some (Negritos, Andaman Islanders) not sharing in the least those characteristics of the Caucasoid-like Australids, (Papuans, Ainus, Polynesians), the genetic infusions may have been few and scattered. Considering the small population, however, the infusions may have been enough so that the characteristics were passed down and are still faintly recognizable.

Not so the Mongoloids. One interesting dimension is the fact that unlike the other races that are color variable—Caucasoids, for example, vary from bleached Nordics to brown-black Bengalis—the Mongoloids are much the same color, whether they are northern Mongols or equatorial Jibaros in the Amazon valley.

The gene for color adaptation to geographic and climatic conditions is apparently dominant; as evidence for this, it is so widely distributed among the races. The test of the Mongoloid relationship often hinges on this unique color factor. The sallow, yellow-brown color of Mongoloids, since it is similar to the Bushmanoids of Africa, as noted in the last chapter, suggests the possibility of an ancient connection between these two groups. The color similarity is doubly intriguing, since the Bushmen inhabit a tropical climate. These are, of course, only conjectures.

Even more interesting are the seventeen basic structural features characteristic of Mongoloids (a few paleoarctic peoples of northern Europe excepted). These features can be traced back to the earliest levels of the Choukoutien excavations south of Peking of *Homo erectus pekinensis*, 500,000 B.P. They are still to be found in the modern northeast Mongoloids today.

Listing only a few will show that they are typical of the secondary physical characteristics that separate the races: "Inca" bones (separated from occipital) in the skull, broad nasal bones, nasal saddle, shovel incisors, external growth on the border of the tympanic plate opening (like infant's) on top of skull. These are not especially crucial structures, adaptively; thus they would likely be retained over many thousands of generations. Why this stable uniqueness?

Consider, if these typical erectine characteristics were already in place 500,000 years ago, one could argue that the northern continental erectines around Peking (which *is* far north in China) had already been isolated for many thousands of years. Since some *Homo* erectines left Africa

1,500,000 to 1,000,000 B.P., these Peking erectines could have been completely isolated for several hundred thousand years. During that time, random variations in a number of nonselective characteristics could have fixed themselves in these populations as secondary racial features.

That these features have maintained themselves throughout the course of the Mongoloids' long history shows the tremendous tenacity of such secondary characteristics and the stable population dynamics within the Mongoloid middle kingdom. Another look at a cultural-historical map of China will show that the earliest historical civilization in China, even before the Shang period of 2000 to 1500 B.C., is centered around this Yellow River homeland of north China. Some paleoarcheologists even question whether or not the widespread Acheulean technology accompanied the Mongoloid erectines to their homeland. Others think its absence is only because the materials in that area were not amenable to this technology. The latter view needs to be made more persuasive. The overall isolation of the Peking erectines would argue against it.

By the end of Pleistocene 30,000 to 15,000 years ago, the Upper Choukoutien cave fossil record shows that highly evolved sapient Mongoloids inhabited this area. At the same time there is evidence of a first migration south and east out of the homeland into the southeast Asian turf of the Australid *Homo* erectines. It is from this period that the hybrid peoples of the East Indies (Indonesians), Philippines, Malaysia, Cambodia began to be formed.

Another mysterious, apparently hybrid Mongoloid set of peoples crossed over the Bering Strait into North America. At roughly the same time, the "pure-blooded" Australian aborigines were now pushed out of their island homeland to the north to occupy Australia and Tasmania. In the eastern islands also, mankind was on the move.

Unquestionably when the Mongoloids started their move they were in full control. The Australids had alway been behind the Mongoloids in endocranial capacity. What we have here is merely the typical historical advance of the northerners, the final morphological and cultural product of more powerful selective pressures acting on a dynamic of variation for brain size and modern structure.

## Intelligence Selects

A good scenario can be imagined for the rather fixed and isolated domicile of Peking man and his descendants once they arrived on the north Chinese plain. Harry Shapiro, writing about *Homo erectus pekinensis'* life, conjectures that the various tribal families probably staked out fifty-square-mile areas within which they attempted to supply their food and shelter needs: berries, wild fruit, game, and a series of shelters in which to make fire and renew themselves for their limited explorations.

Interestingly, though the migratory urge is powerful in man, some tribal members will prefer to stick close to home to keep the hearth warm and be ready to welcome back the wanderers. It is an ancient human tradition and will probably never change much. Early Peking man seems to have remained stable geographically.

Let us examine another curious enigma in the story of the Mongoloids. These Upper Choukoutien cave Mongoloids about which we spoke above were contemporaries of the Cro-Magnon *Homo sapiens sapiens* of Europe. These Mongoloids also gave some evidence of being progressively hominized sapiens. They had been victims of a massacre. All the skulls from this cave had been thoroughly mashed, intentionally so. Associated with these and other late-Pleistocene fossils was a stone technology that was basically only a hair more modernized than the erectine chopper tradition. Where was the rich, delicate, creative contemporary technology of the Cro-Magnons?

Until now at least, we have no record that these northeast

Mongoloids produced anything to rival that of the European cave civilization far to the west. How could these sapients have arisen without the development of an advanced technology? We answer the question with an old question: which came first, the chicken or the egg?

Our basic hypothesis comes in for more use here. Probably little contact occurred between bands at that point in the ferociously severe Ice Age climates, and what contact occurred was not extensive enough to transmit cultural technology and consequently jar the Orientals into a different way of life. (Witness the difficulty with which both nineteenth-century Japan and twentieth-century China moved from their ways.)

On the other hand, on the basis of the present fossil record (no erectine-sapient borderline fossils having been found), we are prevented from postulating a continuous line of progress in China. It is almost as if the curtain closed on *Homo erectus pekinensis* some 300,000 years B.P. When it opened again (not until about 30,000 B.P.), with the same scenery, the actors had become fully-evolved generalized Mongoloid sapients, *Homo sapiens sapiens*.

Could it be that *Homo sapiens sapiens* genes came in bits and pieces from the west, from remnant individuals or groups of struggling, wandering dissidents whose genetic infusions were without cultural power, and whose genes were unobtrusively absorbed into the larger mass of Mongoloid individuals? That would be my guess. In the north the selective pressures were great and continuous. Large endocranial erectine examples (1200 $cm^3$) were common in northeast Asia from as early as 300,000 B.P. The foreign *Homo sapiens* genes would easily meld with the older eastern erectine model and selectively form a wholly new, but different, *Homo sapiens sapiens* type.

Here was a highly advanced physical type of sapient without the technology and art that might be expected to go with it. How was this possible? We usually think of tools as being the keys to human survival and advance. Here again, with our traditional thinking, we have put the cart before the horse.

Tools were the product of an advanced brain, not vice versa. As pointed out earlier (in Chapter XI), the mutations for larger cranial structures and a modernized sapient brain were produced autonomously as part of a hoary hominid evolutionary tradition. All that was needed from tools was, so to speak, that they aid in the selection of the most intelligent. The difference between the most intelligent and the least intelligent in any interbreeding group was more likely to lay in the manner in which *any* tool was used rather than in the availability of a particular advanced tool kit.

Simply, the developed erectine tools that Upper Choukoutien sapients used were good enough to differentiate culturally the existing variations for higher intelligence. These individuals used their brains more than they used their tools and the best reproduced themselves more efficiently. At any rate, the tools of Cro-Magnon were often so delicately crafted as to be of little practical use in the rush and violence of the hunt and war.

There is a similarity here to our contemporary situation. We are creating sophisticated forms of warfare—planes, tanks, electronically controlled equipment—that go far beyond the practical, often gut-response, needs of our warriors. Tools often reflect intelligence, but we ought not assume that their sophistication or lack thereof constitues the key to the particular selective mystery. Give me smarter warriors, no matter what the available tools are.

## Evolving Beyond the Ice Age

Today, we note the Mongoloid influence all the way west to Afghanistan, Persia, Turkey, and Russia. The highways that once directed migration eastward now carry it west—comparable to the Negroid impact on the Caucasoid north—

which has been evident now for at least a thousand years
Highways of human migrations rarely remain one-way
streets.

The Mongoloids today are among the most sapientized
esthetically sensitive, intellectually competent of hominid
forms. How were they transformed from culturally rough,
merely able northen sapients, faintly scratched with the
racial tatoo of an ancient erectine type, into the most
challenging peoples of our modern world?

As late as the Shang dynasty, ca. 1500 B.C., their culture
was still without any hint of the beauty that can be noted in
the ceramic, stone, and metal work of the later dynasties.
Their large copper bowls were competent and functional yet
without the delicacy that characterizes subsequent Chinese,
Japanese, and Korean art. Of course there was the culture of
northern Thailand, possibly contemporary or even prior to
that of ancient Sumer, ca. 3000 B.C. The Shang conquerors
(2000 B.C.) of the indigenous Yellow River peoples were
horsemen. Shortly after their arrival from the west came the
stimulation for the evolution of the modern Chinese state.
Writing was soon in use and the process of cultural ferment
well on its way.

Carleton Coon argues for this west-to-east cultural push
despite the lack of genetic evidence that might support such
a hypothesis. (The Chinese seem rock steady in their skeletal
features to our own day.) C. D. Darlington also speculates on
this, stating that despite great Arabic trading influence in
Amoy and Canton during the Middle Ages, these Arabic
enclaves seem to have left no lasting genetic mark on the
people, though there is historical evidence of interbreeding.

The Chinese probably have changed, for they have
exposed themselves to a relentless process of internal
selection, even into the historical period. My guess is that the
modern northeast Asiatics, literally under our eyes, have
been galloping into the modern world gradually absorbing
the best of what they have received from the outside. They

have accepted these influences but shaped them into a biological type that is probably psychologically and culturally integral to their ancient traditions. Indeed, the Chinese may be different from us westerners. However, they and their Korean and Japanese brethren have struggled through their primeval and traditional problems to create a human type that is now at the threshold of opening a door into our next century.

They can look back to Peking man in all his gross primitivity and say with pride, "Yes, he is our ancestor erectine." They have gone a long way since. I can almost hear their conversations in Peking, relating to their fascinated Caucasoid guests, "Cro-Magnon was your peak. He is testimony to the faith you Europeans kept with evolutionary history. But you let up, gradually at first. Now your sun is rapidly fading. We are ready to take up the hominid destiny!"

# XVI

# Ultimate Sapiens: Cro-Magnon

**Here We See Human Nature**

Neither you nor I can imagine what 25,000 years mean in the life, experiences, and memories of a people. Can it be that such a length of time, during which both history and nature reiterate certain similar themes of life and survival, would not indelibly etch its shaping message into the souls of such peoples? One thousand generations came and went on those bare tundras of southern Europe during which Cro-Magnon or *Homo sapiens sapiens* holed up in the protective caves formerly inhabited by who knows how many generations of transitional humans.

If ever there was a moment in the evolution of life for a complex, evolving creature to show us its animal nature, it was then. In Cro-Magnon we find a clear message of arrival, without the doubts and enigmas of the other, then contemporary, hominids. The human form, in this frosty northern incubator, had received, if not its perfection, its uneasy complex of permanent components. Here man peaked—in brain size, hominization, in naive esthetic exuberance, spontaneous sexual overflow, in an abandon, sometimes an orgy, of energetic aggression against human fellows and animal prey. Still he wandered. We trace the ebb and flow of cultures over 25,000 years, phantom gray shadows across a blue-white horizon, imprinting their humanness on those deep inner cave walls, ceilings, even floors. In the south there was obviously other movement. The world was in process. In slow trickles the shapes of sapiency reached out to the southern lands beyond. Cro-Magnon was the first truly *Homo sapiens sapiens.*

Let us examine the significance of Cro-Magnon for ourselves, both what he represents biologically and as a creature of society and culture. Because he came onto the scene so suddenly and began to show his stuff without any outside prompting, we can take seriously the claim for a causal relationship between what he was as a biological creature and what he produced in his social life.

## A Creative Life

What we are of course is an accident of brain size and structure. What nature gave us by chance, in varied shapes and sizes, we had to take and use, if possible. That this big baby-faced creature fell heir to such an extraordinary cranial protuberance was one of those miracles of time and evolutionary destiny. For nature, it was an ordinary aspect of her repertoire of adaptive possibilities. To us, who wrestle with ourselves, it may seem miraculous. Nature itself recedes into the background and we humans, ever more conscious of our power and uniqueness, tend to stand apart. With our suzerainty over the rest of nature, at least our apparent suzerainty, we tend toward self-indulgence. We forget that a biological leash is attached to us. At some point we will pull too hard and it will turn us over.

Cro-Magnon did not have that worry yet. Wandering forth from his ancestral homeland seeking a better life, he was naturally drawn to the teeming tundras. The killing and the food were easy. As we might, in our own day, meeting only prosperity, knowing little of the spindle of necessity, he too gorged himself and energetically shaped his life around what he hoped was an unlimited bonanza.

Where Neanderthal worked hard at life—indeed he made a decent living—Cro-Magnon turned life into a cornucopia. We see on the outside of caves long deep ditches filled with the bones of food animals. Indeed, even there, a community dump. Eventually, the mastodon and mammoth were killed off, as they were later by the Amero-Indians of the New World. Other, less delicate creatures suffered the agony of human pursuit and the kill with less harmful results. By the end of the Pleistocene, for example, there is evidence that Cro-Magnon was herding reindeer, culling out a limited number each year so as to maintain a long-term and steady supply of meat and animal parts. He was beginning to learn. Had it even then become a matter of necessity?

Distinguish if you will between ignorance and stupidity. Ignorance is lack of knowledge (remediable); stupidity is lack of the ability to utilize knowledge (for the most part irremediable). Cro-Magnon was as big-brained and as sapientized a type as any of our contemporary *Homo sapiens sapiens*. Thus it would be wrong to compare him with the natives that his European descendants found a century or so ago. This error is comparable to our using contemporary anthropoids as a model of ancient hominid behavior.

For one, Cro-Magnon was not cooped up on a restricted island or in a well-carved-up jungle sanctuary. The evidence is that he wandered across the breadth of the Eurasian steppes. Yet because of the climatic conditions the only time he could go any distance was in the short, cool summers; as Romuald Schild postulates, at this time, especially toward the very end of the period, the tribal groups apparently met, to exchange goods, take wives.

The overall impression of *Homo sapiens sapiens*, as compared with earlier hominids, is of a cultural creativity that takes off like a rocket. Was this nature's way of heralding the completion of *Homo*'s morphological transformation? Did the Cro-Magnon morphology test the limits of nature's ability—for instance as it widened the bipedal sapiens female pelvis to accommodate the birth of big-headed babies?

The archeologists have identified Cro-Magnon's cultural traditions and traced them over an area of almost four thousand miles, from deep in Spain to the Ural Mountains at the gateway to Siberian Asia. Certain traditions in technology, in the arts, painting, sculpture and in the choice of available materials limited as well as enhanced the power of Cro-Magnon's creative genius. The names Aurignacian, Gravettian, Solutrean, and the final icy encapsulement of the Magdalenian antler culture identify the traditions.

The Cro-Magnons had personality. It is almost possible even to recognize an artist's work, to recognize an individual woman's face in the shaped bone. What we are seeing are cultural traditions, as tribes wandered from place to place and time to time leaving their cultural handiwork behind, sometimes even returning to a place several thousand years later. We must remember that the progress in culture—and interestingly, it is progress—occurred as the tools overcame the material intransigence of the rock, the hematite, the ivory. What we see is more than a delicate punch blade. We see an overflowing brain.

## Esthetics, Technology, and Sensuality

A new technology had been born. Instead of the roughly-beautiful Acheulean hand ax or the crude flint strikes off a tortoise rock core of the Neanderthal, a whole series of delicate double-edged flint tools was now being produced. The so-called laurel-leaf blades of the Solutreans were delicately-edged tools with little scallops removed from both sides, reminiscent of laurel leaves. Some of the Solutrean tools were obviously made as models or as delicious experiments. So fragile were they that they could not possibly have served as practical instruments for home, hunting, or war use. The bow and arrow were now in use, as was the spear thrower, the latter often decorated with carvings—as a medieval knight might decorate his sword or an American frontiersman the wooden stock of his rifle.

The technology, as awesomely beautiful as it is, almost pales beside the larger than life cave paintings and sculptings discovered at Lascaux in southwestern France and the later Magdalenian period frescoes discovered in Altamira in Spain. These decorated rooms deep inside caves remind us now of our cathedrals; perhaps they served as a kind of subliminal example for the Sistine Chapel. They probably were sacred rooms uniting the religious and the esthetic.

In one painting a man was represented dressed as an animal; deer horns decorated his headdress. He was obviously dancing in some totemistic ritual evocation. We wonder about the function of this world of art and religion. Human figures were rare in these paintings. More common were drawings and paintings of the animals that surrounded the Cro-Magnons, the animals on which they were dependent.

I am skeptical that these paintings were used to conjure up the images of hunt animals so as to empower the hunter in stalking his prey. In fact we ought to move beyond any literalistic or magical view of what the Cro-Magnons were up to. They had 28,000 years to figure out their environment within the constraints of the climate around them. True, lack of experience and depth of geographical perspective prevented them from knowing what else was possible or how else they could live. That would come later, when the ice receded and they were forced to descend into the southern valleys to learn new trades and a new way of life.

They had to be practical when practical meant surviving. Their relative knowledge was great, but one mistake would reveal the depths of their ignorance. On the other hand, those neural energies poured through the billions of brain cells into their entire psychophysiological system. There was body as well as brain, a lower, mammalian brain as well as cerebral cortex, just as today. Men and women rejoiced in shows, festivities. Just as we include in our own celebrations the decoration and hoopla, the parades and floats, airplanes swooping overhead, or the exciting halftime entertainments of the football games, Cro-Magnon man was equally festive.

The measure of the people and their culture lay in how they handled those basic mammalian-anthropoid yearnings and channeled them through the cerebral cortex. The beauty, subtlety, and abstraction of the painting and the sculpting have invited much analysis. Little new could be added to these analyses except that the ancient human sense of artistic form was here welded to new techniques of paint making and application, to sculpting, and to the discovery of materials to produce creations of higher intellectual value.

Here, in this first civilization, we see what a little prosperity will do if the people have the discipline to match their intellectual abilities. Not having television to watch or cars to take them to the corner pizza parlor, they entertained themselves in the dark recesses of their caves, evoking and objectifying their world of meanings. At the high point of their creative and intellectual power, they revealed to us the relationship between material plenty and the release of this heretofore unknown symbolic capability lurking in Cro-Magnon's reorganized hominid brain.

We wonder at the many Venuses found in the archeological sites inhabited by the Cro-Magnon peoples. The Venuses are the first sculptures found of humans. They are female subjects, obviously sculpted by male craftsmen. In all of the nude figures is a moving ripeness and voluptuousness. Rarely were features of the face carved out. The head of the lady of Brassempouy is an exception, the delicate features revealing a loving rendering of a human being one would like to meet. Delicate, even features, perhaps like the lady next door, yet this was created 25,000 years ago.

Others, ample bosoms folding over pregnant-appearing torsos, large, often overmatured *derrieres*, in one case a vulva as a separate piece of sculpture, point to a powerful and magnetic sexuality. Some archeologists have argued that these were fertility symbols. These were a people for whom life was a constant effort to sustain. Women who could bear children were admired, possibly even revered. Balanced against the fertility of the human female was an intense awareness of the fruits of fertility in the animal and living world in general.

The Cro-Magnons were successful and comfortable humans enjoying much prosperity. One's fertility could be as great a problem as it is today. Really, are we to believe that these smart humans did not know the facts of human fertility: pregnancy, conception, its causes and sequences? Would they worship and revere something so natural to mankind, so understandable, even controllable?

I would offer another explanation, one that might incorporate the fertility idea, for it does adhere to the female principle and is probably never completely out of mind when males interest themselves in the female. My bet is that the key to the problem is not merely an overly imaginative creative intellectual thrust. The Cro-Magnons were not too different from us moderns when we are at our best. With plenty of meat protein, plenty of time on their hands, and no television, they may have had something much simpler in mind—sex.

We have *Playboy*, the Hindus had their temples at Khajraho in Madhya Pradesh, the Greeks their cult of Aphrodite and innumerable other semireligious sexual rituals. The Cro-Magnons created sculptings, many of which were mobiliary (carried) objects, carved smooth, hard, cold but warmed by the hand. These soothing objects called up delicious memories; perhaps they were carved while the men were out on the hunt. They were artistic reminders, as they are today for men away from home, of beauty, sexuality, fun, warmth on a cold night; they were objectivizations of the powerful surges of energy represented in the lustful possession of a round, bubbling pulchritudinous female body.

Sensuality demands a symbolic reminder. To the curious, wondering, overly intelligent and energetic but naive Cro-Magnon, it all must have been tumultuous, even adolescent. Most probably tough rules and regulations reined in his volcanic energy and passion, channeling them into socially useful directions. Art was probably one way that all that lust could be worked out of the biopsychological system or profitably sublimated.

## Humor=Civilization

What else went on in their day-to-day, year-to-year rhythm of life we cannot know. Surely the Cro-Magnons must have examined the world with curiosity and wonder, at least to try to fathom its workings when outside events threatened their survival. The hints that we get from observing the externals of their lives—tools, garbage, artifacts, painting, show this new intelligence attempting to grasp the world from the outside, to subject it to ordering principles: principles of life and energy, birth and death, of sexuality and fecundity, of passion, sleep, and renewal, of ritual, order, significance, and sociality, even of playfulness and humor.

There is a spear thrower that comes down from the late Magdalenian era 12,000 to 10,000 years ago. It is made of reindeer bone. The top of the shaft is carved in the shape of a frolicsome fawn. Its rear end is raised up in a familiar position—it is defecating—and the head of the fawn is twisted as it half-examines the event. The droppings are on the way out and down. The fawn's tail is shaped into a hook; the droppings themselves are fixed to establish the spear shaft. It is an ingenious creation. What scatalogical mind could envision in a spear thrower that noble beast in the midst of a typical if clumsy natural act?

When the spear thrower was finished and put to use, it probably got as many admiring laughs from the fellows on the hunt as it would today. Where there is humor, there is the ability to rationalize and order experience, to satirize or joke about it. One can stand off from the mire of instinct's rigid behavioral demands. Humor tells us of man's freedom to think and act.

# XVII

## Thinking Man

### Abstraction

The stars are bright over a New England hillside. The cloud of sparkles, parted by a puff of haze, scatters the sky with tiny diamonds of light. I wonder, probably as did Cro-Magnon, how odd, how mysterious, that one would even gaze up and take notice. From my hillside perch, I can follow the wanderers, the planets, as they dip through the maples and disappear below the western ridge. The moon edges up and over in the east on its varying yet repetitive passes through the sky.

Cro-Magnon man saw it all too. The regularity and variability in this evening drama must have set his mind buzzing. What did it mean? How did these images relate to the events in his life that appeared and disappeared with a rhythm that needed to be caught, managed and at least intellectually noted. A nick in a piece of wood, a diagonal slash through four verticals. That's what we do today, but only after ten millennia of accumulated knowledge. What did they do in the ten or fifteen millennia before that?

Alexander Marshack's *Roots of Civilization* and his many other writings since have explored the meaning of some heretofore disregarded bits of evidence about the life and mind of Cro-Magnon. What had been thought to be series of quasi-geometric designs or doodlings in the wall paintings or on the various artifacts were reexamined. Sometimes on the cool walls, sometimes on a piece of stone or flint, bone or ivory were found regular patterns of incised markings. Some of these incisions were done on incredibly small pieces of stone or bone, at times with what seem to have been different instruments. Sometimes, even on the same plaque, various patterns of incision could be found.

Marshack, a former magazine editor, at the suggestion of Hallam L. Movius, Jr. of Harvard's Peabody Museum, began studying these "notational patterns." The outcome was a totally different interpretation from that assumed by generations of anthropologists. To Marshack the occasionally serpentine swirls of incisions were the product of man trying to fathom the events of his world through an outside objective instrument of thought. By making a mark we in a sense visually prime our memory. A mark stands for an event or an idea.

There is no sense in making marks if each one stands for a different category of things. To be useful, the marks must be part of a system, as are the flags on the side of a fighter plane indicating the number of airplanes the pilot has shot down, or the bombs painted on a bomber indicating how many missions the plane has flown. Notice that in each case the symbols, "flag" and "bomb," were not literal pictures; rather they were symbols representing ideas that you and I would recognize. A flag in this case was not just any flag but a plane of that symbolized nation that had been met and downed. The bomb did not represent the quantity of bombs dropped, rather a successful mission accomplished.

Of course, some relationship should exist between symbol and meaning; it cannot be completely arbitrary. It would be odd if the fighter pilot plastered fifteen decals of Mickey Mouse on his plane or if the bomber sported pictures of cottages with white picket fences. The symbol has to be reasonably meaningful.

Marshack found the symbols on the walls even alongside the pictorial panorama of animals, plants. On occasion he noted a series of animals, perhaps representing several hunts, then another with an animal or animals only barely sketched, which Marshack interpreted to mean "and so on" (etc.).

The notations or markings were different. The cave men were not necessarily recording events, but some kind of abstract numerical function, perhaps the relationship of time variables or different events, such as the phases of the moon and the time or length of a woman's pregnancy, of the days between her menstrual periods. When the skies were clouded—which would prevent complete visual or experiential accuracy—or when a tribesman might be away on the hunt, the plaques (some were easily carried) were convenient abaci to record numerical units and thus keep the events solidly in phase.

By so notching his material, the early philosopher could check if the same number of notches appeared for each woman's pregnancy or menstrual period, or for the relationship between the moon and the seasons, or for the coming of the first swallows. Naturally, the incisions could go on forever. We ourselves use a simple system: ⊔⊔⊓ 111, for example, is an easy shorthand way of noting a sequence of five plus three, whether objects, days, or yards. Note that the things we categorize can vary, but the category of the concept, "5" or "8," stays the same.

This ⊔⊔⊓ =5 was an extremely sophisticated discovery, which, unfortunately for Cro-Magnon, was not available to him. He had to discover these ideas for himself. How great was his intellectual mastery of experience—of the mile, the foot, the year, the month, the seasons, gestation periods in animals and man, migration patterns, larger, smaller, far, close, many, few—we cannot know for sure. Even the idea of left and right, so basic to us, might have been unrecognized by him. We see in the carving of incisions the discontinuation of one marking instrument and a beginning with another. What did it mean, since there is no break in the markings, just a new point and shape cutting into the material?

We get some sense of the relativity of things when we see Jews reading right to left, and Chinese top to bottom, when we work with arithmetic bases different from 10, as in the new math. Marshack is emphatic that these notations show that Cro-Magnon was toying with the concept of a "system" through which each individual is related to others by a principle of association, just as in our ordinal system of numbers, 1,2,3,4 means 1+1=2, 2+1=3, 3+1=4, etc.

Obviously the success of any new recipe is in the eating. The Cro-Magnons clearly were successful. That is why they had time to do this "intellectual" doodling. We really have the basis here for a civilization—the leisure, the wealth, the comfort, above all the stability over a period of time for the mind to go meandering.

Yet we understand why it did not go further, why a system of notation was not incorporated into the life of the community as were writing, surveying, and accounting in the great valley civilizations a few thousand years hence in the neothermal period. The scale of life was just too small for a highly abstract system of ideas like a written language or number system to be useful or practical. Communities must have inhabited one cave for hundreds if not thousands of years. The localization of life, save intercourse with other groups during the summer, did not demand anything so complex.

As highly intelligent as they were, they probably counted on poets and storytellers to remind them of history and events. Such abstract symbol systems would come into their own only in the complexity of the modern city-state. Still the Cro-Magnons searched for a principle by which the steady, unchanging events and rhythms of their lives could be understood.

## Intelligence Creates Civilization

The Cro-Magnon peoples were the disseminators of the biological possibility of complex civilization. If it were not their particular genes that were spreading out of Europe and western Asia before and during their known residence in the caves, then those genes for sapiency came from people very closely allied. The Cro-Magnons represented in their intelligence these possibilities of civilization for the people who later settled into the river valleys, where the great city-states began.

The Cro-Magnons were capable of that highest dimension of intelligent behavior, the capacity for abstract thought. They were not limited by a mental range that only dealt with concrete things, momentary and unrelated events, and a world close to the bodily and personal needs of individuals.

Claude Lévi-Strauss, in his various writings on the logic of primitive intelligence, properly concentrates on the concrete attitude of primitive peoples. He gleans from their world of particular things the richness and even the routine repetition of events and behavior. Surely there is logic here. As with all primitives, however, whether they inhabit an isolated jungle enclave or live close to a megalopolis, "thinking" is initiated by random events. No originality, no conceptual world exists beyond the immediate individual needs of the primitive thinking person.

No, the great burst of civilization that came later was made possible by the huge sapient brain that nature had miraculously produced. This brain completely overran the last vestiges of primitive mammalian patterns that still enclosed the anthropoids within the bosom of nature. Instinct and embeddedness were now dissolved. Man was forced to live in a new, open, and suddenly dangerous world.

Yet the intelligence that was just marking time in these

Ice age sapients was still in the great tradition of intelligence as a vital adaptation for animal life. Intelligence had always been the riskier road. Intelligence that functions as a guidance system for living things, so to speak, between and beyond the instinctual regularities of the genes, gives the animals some choice for behavior. The key word is "choice."

The amoeba floating in the seas has precious little learning capacity. Perhaps its reaction time will quicken— positively to the sun, negatively to excessively alkaline water. The first time, the amoeba moves in 3 seconds; by the fifth exposure, the amoeba has it down to 2.5 seconds reaction time. It has "learned." That .5 second increase in response speed marks the breath of air injected by learning into rigidly stereotyped behavior.

The condor soaring above the Sierras sees much but reacts to only a few visual stimuli. Much of such decision-making choice is supervised closely by instinctual mechanisms; behavior and intelligence are always linked to clear-cut survival needs. No dancing "ring-around-a-rosy" for animal life. Except for a few youthful moments, when nature allows for play as a means of anticipating possibly new environmental situations, the animal, high or low intelligence, is all business.

The function of intelligence in animals higher on the taxonomic scale of life, from the prudent amoeba to the silly chimpanzee, is to predict events that cannot easily be programmed genetically. The higher on the scale an animal is, the more we find movement, exploration, even curiosity as the creature traverses wider and wider arcs of experience, learning how to make it in the world, anticipating changes, going off in new and opportune directions.

Intelligence frees animal life from the rigid, stereotyped simplicities of surviving. Nature can accept a wanderer or an explorer, because nature is always renewing itself. The past is not always prologue to the present or future. Intelligence is supple, perhaps not powerful, huge, ferocious or even fecund. Intelligence points the creature toward the new and the possible. Thus, intelligence is a practical adaptation. It must meet the needs of those ancient genetic survival strategies. Even man is here absorbed into the sociobiological nexus. He too must be practical.

In the meantime, something awesome as well as worrisome happened. The new, hypertrophic cortex has restructured all previous relationships between intelligence and behavior. Extra brain cells beyond any dream of practical utility—thousands of millions of them—have rushed into the soaring cranial vault and have literally taken over the action. The heavy weight of this electronic existence, merely pulsating in frenetic energetic excitation, dictates a new scenario for the human scene.

Even Neanderthal, a sapient creature of unusually extensive brain structure and intelligence, was a placid live-and-let-live hominid. In many ways Neanderthal society was probably similar to the various nineteenth-century primitive societies that Europeans discovered in which the rhythms of life, thought, and action extended themselves calmly to the world of nature around them. There was no plethora of human energy in evidence. Neanderthals and, indeed, undeveloped modern peoples seem content to live within the boundaries and to harmonize with nature.

Not so modern *Homo sapiens sapiens*. He was aprowl with energy and schemes. As has been said about the nineteenth-century capitalist entrepreneur, he was "calculating." Cro-Magnon was too, but in his ignorance, as in our own, he often did not know his own energies.

Our picture of creative, energetic man does not end with the secession of the ice floes. That was merely a jolt, a temporary setback while man readjusted his mind and his society to a wholly new ecology. The event radically revised demands on his adaptability, but also presented new opportunities: agriculture, urban life. Then the prowling started anew, not to end, as yet.

What resulted was this: What little instinctual restraint and security that remained with the hominids was dissolved in Cro-Magnon and modern man. That welter of nervous energy now generated on its own. The product— unstoppable, only capable of being channeled or, as Freud once put it, in relation to the libido, of being sublimated in new directions—was thought. *Sapiens* intelligence started out as an aid to solving the practical problems posed by the traditional needs of living—hunting, gathering, and defense of hearth and home from other roaming hominids.

The hidden potential bubbled forth. It flowed into an intense and raucous sexuality, showing itself not only in nonstop pursuit of the male or female object, but in sensuality, as in the carved nude, the pictured image, the love game. Ritual, even religion became another means by which this roaring flood of mental eroticism was expressed and contained. Sensuality's basic focus may have been the procreation of children. *Homo sapiens sapiens,* however, somehow reconnected his cortex to his gonads; the electrical contact was shocking.

Finally, we observe the mind/hand scratching out markings day by day to subject the experiences of life to some orderly pattern. Later, the Athenian philosopher Plato and his school would suggest that knowledge is not knowledge until it is put into some kind of mathematical form. The world is too complicated, too dynamic, to be dealt with just as things come along. Without a picture of the inner structure or meaning of even day-to-day life, we would resemble those people who wake up each morning confusedly to a wholly new world.

Human experience, rich, varied, challenging, can no longer be dealt with successfully unless we can discover a deeper web of practical meaning than that which our forefathers left us. We can no longer rely on their solutions. Now we create our own world of relationships that demands new responses from moment to moment. Civilization's mandate requires that we now live in a world of abstract meanings, using a map of predictable events made possible by the fact that we can externalize our thinking, by speaking about the world, by even drawing a diagram on the ground or on a wall. The marks now stand for abstract quantities; they are aids to the mind to organize a vastly greater domain than we could if all our knowledge were enclosed within our skull.

Only when ancient man could extend the physical domain of his life experiences did these mental energies emerge to make their mark on human life and the destiny of all other living things. In essence, this is the difference between the highly sapientized hominids and their less advanced brethren, a difference that then reshaped the history of all creatures.

The logic of the concrete mentality that evolved in those ancient eons is of little use now in dealing with a modern world of computers, corporate development, microchips, the esoterics of mathematics, philosophy, or literature. The heritage of hominid intelligence as applied to the problems of organic adaptation and survival has here, in this one blink of time, found its fulfillment.

# XVIII

# Intelligence at War with Human Nature

**A Parable**

The four men gazed into the distance beyond the parapet, toward the green banks of the canal bed. The tents—there seemed to be more of them this morning—were scattered over the entire canal area. On the horizon figures could be seen using the water that snaked along the edges of the canal. The oldest of the four, tall and grizzled, nodded in that direction. "You think there are more beyond the hills?" A younger man, shorter, wearing a similar shirt and military leather belt, replied: "Our boys think so. They are short, dark, thin-faced people. They waved the scout off. The traders think they remember them from last year when we opened the walls after the harvest. Their language was unintelligible." "How long will it be until the harvest is in this time?" the first asked the third, also short, lean, with a leathery farmer face. "It's big this year. It'll take another four days. I'm afraid we shouldn't wait. We have enough food for all year now, if the priests can ration themselves."

The tall old soldier sighed. "We have talked to the captain twice. The priests have him in their hands. He is fearful of the Gods. He wants to wait until the harvest is in and the libations drunk. Then, after the orgy to the Powers has begun, we can break the dam and flood them out." "It will take our people twice as long to undermine the dam when they're drunk," the fourth man laughed ironically. "We need to get started on it now. Then, in a night it will be ready to collapse with a push. But my men aren't enthusiastic about thinning out the walls now—the crops, the festival, the fun. . . ."

"It will be too late then," the younger soldier observed. "Somehow these people know our habits. We need to flood the canal now. Then, we could come down on them. The survivors would run back to the desert," the older commented. "The people are mad with eagerness for the orgy," the agriculturist replied, "and the priests have them frightened stiff. They won't stop now, not right in the middle of things." "Even if we have some time and get the men on the walls, they will be too tired, drunk, or horny to fight," said the older soldier. "The desert men will get lots of grain, maybe even our city. Why can't we get the council to reconsider?" asked the younger soldier. "All they talk about are the greater powers in the universe, the protection of Gods, and the ancient holiness of our city," the engineer replied. "By the time we try to release the water, the outsiders will be all over us. It will be futile."

The old soldier nodded quietly as he turned away and looked again, beyond, to the hazy, dark mounds in the distance. An array of tents. What was inside each? "We have enough men and weapons," he muttered hopelessly, "but too much fat, too much sleep, stupefying dreams. . . .we will learn the hard way. . . ."

**Ancient Drives**

In every community are a few whose imaginative vision goes beyond the expectations of the mass. This vision consists of a map of possibilities built on the reality of today, but always ready to extend itself and adapt to the probabilities of tomorrow.

The difference between such people and religious visionaries or seers is that the realistic dreamer's map is based on a theoretical picture drawn from real secular things and events. Thus the military adviser and his three intelligent associates can envision the possibility of an attack on a traditionally impregnable city, a city at comfort and peace with itself, but at the same time made susceptible to danger by the incense of theology and ritual that appeals to deeply fundamental human needs. Often reason, prudence, or self-restraint cannot dampen the electrical fires of the contradictory desires that stem from those root mammalian needs and drives.

The vignette described above is only one of innumerable possible situations that exemplify what humans have run across from the time of recorded history onward. In the late twentieth century, one enormous and real situation confronting us is the contrast between the fact of proliferating populations, dwindling resources of land, food, and energy, and the fact that we now have the ability to limit these populations through birth control. The social consensus for a good life built upon a balance between population, resources, and the quality of things that we want to create is short-circuited by the inertia of religious, moral, and other mental and emotional blockades that militates against achieving our ends. We are at war with ourselves. The wars are sometimes concerned with material issues, more often with the ethnic, religious, national, ideological. Internal dogmatic, emotional, or symbolic barriers prevent us from achieving the practical ends that could make the difference between survival and obliteration.

The problem is real. It is contemporary. Its sources lie in the precipitous, almost accidental creation of a supersapient brain. Instinct was well on the run even with the coming of the erectines. Of course, intelligence was then limited in power. Man was limited by his instinctless vulnerability, his fragile children, his relatively weak body, the extremes of unmastered drought and cold, even human disease.

H. R. Wallace, the nineteenth-century English naturalist, phrased it accurately. Why did nature create a brain with an endocranial capacity of 1600 cm³ when one of 900 cm³ did work and would have worked adequately for the survival of the species? Unquestionably, as I have explained in an earlier chapter, the final sapient brain was a freak of nature. It was a brain that upset the balances and created wholly new dynamics. These were fully established by the time of the Cro-Magnon cultures. These cultures were only released to grow into full-fledged civilizations when the ice retreated. Man then descended into the river valleys and the shores of the southern seas to create a new way of life under the whip of developing new survival strategies.

All our difficulties as well as our possibilities have come about because the sapient brain overrode the last restraints, indeed the directiveness, that instinct gives to animals. In the animal world intelligence is guided by instinct to achieve clear-cut survival needs. The cat, the dog, the ape, are all intelligent animals. In the wild, their intelligence is enmeshed in a web of practical goals in which their physical bodies as well as their mental energies are closely inter-connected.

Man, the generalized intelligent ape, has a brain that establishes the rules of the game, almost irrespective of the individual's bodily or even grossly survivalistic needs. Man can voluntarily remain celibate, commit suicide, murder his children, act in violation of all the basic laws of natural survival. What remains to man after these basic conservational restraints of instinct are extinguished is a super intelligence that filters all major categories of human behavior through the cortex. He is a thinking animal with a complex brain, a supremely energized mammalian brain that must now control, direct, guide his behavior. The old passions, energies, and drives no longer have built-in censors. Man can go wild in passionate madness or freeze in a catatonic fit.

Let us talk about that ancient brain, that glorious source of humanity, our feelingness, and our closeness to nature. Here we harken back not merely to an intellectual center that was then more limited in size and function, but rather to a mesh of powerful subdivisions that lay under the cortical canopy. These "limbic" structures (thalamus, hypothalamus, amygdala, pituitary, hippocampus), born in the years when the mammals staked out their separateness as a defense against the dinosaur host, grew to spectacular prominence in the millions of years subsequent to the demise of the great reptiles.

Herein lay the passions of motherly protection, the new sexuality and individuality that came with a familial division of labor, the suckling and nurturing of the young, the winning of the mate and the securing of one's genetic immortality through the family. Territoriality is an important part of mammalian protectiveness and exclusivity and the passions that grow from it would one day generate national states out of the tiny bands. Ethnicity, early in the human adventure earned its place in the sun.

The entire repertoire of emotional and passionate feelingness about the world that originated in the mammals became energized beyond recognition in man. In man, it went into a partnership with the new brain. Neuron was piled upon neuron; the consequence was a lightning storm of energies only incompletely directed by the new intelligence. Before, in animals, these powerful drives and emotions were wedded to real, instinctually regulated needs. Now the priest, the sorceress, the magician, the rock star, the cult hero, the *Führer* could become charged with demonic powers that overrode the long-distance good, even the survival of the community.

The four lonely leaders of a city under the thrall of myth and passion could only await their tragic fate along with their fellows. Intelligence had failed here under conditions of newness for which the citizens were unprepared and uneducated to anticipate. Like Plato's lonely navigators, ignored and maligned by the rank and file sailors who listened to the beguiling messages of the demagogues, they could only weep at a human condition that has been with us from time immemorial.

The old questions remain. Can a ship of state be guided by an untutored and susceptible rabble? Is there a knowledge or education that can withstand the enthusiasm, the exciting images of myth and pseudoreality?

Leaders, exploit the old brain and disregard the new, but at your peril! The exhilaration of the ecstatic will inevitably be silenced by the grinding rocks as they tear at the drifting keel of the nation.

## Contending Minds

*Homo* has long had art, language, ritual, technology, religion, music, dance, all the aspects of humanity that we cherish. Most of the evidence of ancient cultures has disappeared. A few fragments, shaped rocks are all that signal to us from a million or so years back. Yet culture was there, even if rudimentary and undeveloped. Then came the eruptive brain and a surging intelligence. All bets were off as the dynamics of change and complexity of the human experience took off like a rocket. Where is it headed? No one can know, for it is nature's latest and most unique creation. We have no blueprints for this brain and its outflow.

Man must make his own destiny. While the new cortical brain may seem to overwhelm the older structures, the older structures are still there, and now infused with the blazing energies of human passion. The circuitry is novel. It violates all the ancient animalian principles. No experience can translate itself completely into a pure and clear mammalian behavioral act whose preordained consequence is to insure the survival of the individual. All experiences must now flow through the cortex or the symbol system, through the filter of human culture.

To the extent that high intelligence is available to guide humans practically and instrumentally, it could work for the good. *Homo*'s expansion and building of great and complex civilizations are testimony to the secular thrust of intelligence to enable man to learn through experience and thus to predict a more secure future. At the same time, intelligence acts to turn mammalian limbic system drives and the lower hominid structures of behavior into the great glories of human intellectual civilization. Defense of the young becomes tenderness and protectiveness for all humans, young and old. The sex drive becomes love and commitment. The frenzy of ritual immolation becomes the sacredness of the marriage ceremony, glorious Fourth of July pageants, Thanksgiving feasts. The shaman becomes the wise priest, rabbi, or canon. The totemistic dance around the fire becomes classical ballet, the recitation of chase and hunt is transformed into Aeschylus or Shakespeare. The pounding drum and the vibrated leaf become the string quartet. What is basic in culture is transformed into ever new cortical forms of behavior with the consequent possibilities of development, accumulation, and recreation. The brain no longer settles into a convenient niche. Its search for meaning reaches beyond the most fundamental and ancient fascinations.

Of course it doesn't always make it. Religion, which in highly intellectual civilizations becomes ethical, often monotheistic, can revert to the shamanistic, to the chants of ignorance and absolutes. It can challenge intelligence for its regulation of human destiny because intelligence is tentative, experimental, always changing. Religion burrows deeply into the basic mammalian feelings of dependency and need for group survival. Touch the right chord at the right moment and you can wipe out a century of careful planning and building.

Witness Nazi Germany. Intelligent, highly educated people, in a moment of crisis rejected reason and prudence for the ancient symbolism of blood and fire. No system of ethical, religious, or philosophical restraint could have prevented the genocide and universal destruction that accompanied this cultural cancer.

Yet little more than a generation later, the fires have been dampened, the cities rebuilt, the shame expurgated. The people, under a new national tutelage have turned away from the ancient expressive system. The Germans are an intelligent people still, despite the bestial sacrifice both of their own best young men and of millions more of the most creative human beings in the world, their purported "enemies." What a waste!

Intelligence is the evolutionary guidance system for the future. The destiny of no animal is foreordained. Our genes no longer determine our behavior nor do they map out the best method of fulfilling our dreams. The evolutionary trade-off lies in our capacity to create platforms of ideas (philosophies) to help guide us into what we hope to be fruitful pathways and directions. At the same time, these are only hypotheses, guesses. Can we ever know for sure if such goals or directions really fit what we are as human beings?

Even trying out a program or plan becomes a dangerous undertaking. Those not intelligent enough to comprehend fully, to plan, program, intuit cause and effect, will be distracted by more emotional, mythological persuasions and scuttle the entire venture. Every such failure of intelligence to do its work efficiently runs the risk of opening wider that Pandora's box of passions and irrationalities. That is why so many societies have fallen into stagnation; they are so frozen by the fears of ignorance and failure that they fall into Pharaonic routine and obedience, one step above the ultimate primitiveness of our presapient ancestors.

Intelligence is the dangerous adaptation. Once we have tasted its freedoms, it cannot be let go. The only way we will ever be liberated from its urgings to create a new method or find a new solution is to surrender thought and learning to the primitiveness of limbic system dominance. On a worldwide scale, that result is probably unlikely. Highly selected now for millions of years, intelligence is a germ whose infectious consequences will not easily be thwarted. It is in us. The war with the ancient past will continue.

# XIX

# Epilogue: Uneasy Victory

In the fall, at first the river ice packs dripped downstream longer than they usually did. Rings of water continued to surround the icebound lakes. At the spring solstice the sun seemed warmer than before. Birds were seen experimenting with the April breezes.

Next, year by year, then decade after decade, alders, larches, poplars and birches began spreading farther and farther north, unevenly along the river banks now full of melt water, then in the upland hills. In barely a century, spruce, pine, and hemlock were appearing far from their usual haunts.

The ice was in retreat, a new interstadial in the offing. After so many thousands of years the warmth was returning and a new human being was being browned in the July heat. Finally, and calamitously, came the hardwoods, beech, oak, maple. Into their shadows the herds disappeared. The whole beautiful scenario collapsed in front of the Cro-Magnons; their sustenance evaporated almost overnight.

The world was still relatively empty, so Cro-Magnon could move out. Some followed the tundras north to the Baltic coast and hung on for dear life. Others moved to the southern coast, existed on mussels, painted the rocks with their multicolored stains of berries and clays, and remembered the chilly but good old days of plenty. Soon in the highlands of west Asia new possibilities appeared, to be honed on necessity and thought.

Encampments from as early as 8000 B.C. hint at animal domestication, the gathering of wild food grains, lithic industries, and the beginnings of new communities, eventually modern civilization. Our present neothermal period continued from that point on some ten thousand years ago. When the cold will return we cannot know for sure. Based on our knowledge of the Pleistocene experience, these warm interstadials are of short duration, perhaps as brief as 10,000 to 25,000 years, in some cases longer. Possibly the ice will hold off beyond the lifetimes of our great-grandchildren.

The events now written into the historical record need little further discussion: urban society, agriculture, metallurgy, the invention of writing, the widening and deepening of the esthetic imagination. The real problem is to explain these events from the standpoint of human evolution.

Mankind proliferated, advanced peoples and their cultures spread in the same concentric circles as in the late Pleistocene some hundred thousand years earlier. Only now it was a veritable inundation, with populations often leapfrogging over great geographical distances to set up their cultural and genetic outposts. The intermixing of this human species, already well underway, continued, so that by the late twentieth century, only in the northeast quadrant of Asia could we say that that most isolated and distinct of all the new post-Pleistocene races of man had been able to maintain its evolutionary integrity, at least for this moment.

Throughout these ten thousand years the gradual accumulation of social experience had precipitated an increasingly intensive dynamic of cultural change. Key to the change was *Homo sapiens*' discovery or invention of crucial instrumentalities that helped him to gain some control over his physical environment. These discoveries took place in those civilized oases that the northern icebound forms of man had established in the valleys and archipelagoes of the south.

Throughout most of these thousands of years. the social scene was still one of great travail as humans struggled to earn their living as well as to create the moments of leisure in which to experience the intellectual delights demanded by the mind. It is fair to say that the avalanche of symbols pouring from this supersapient mind enfiladed every human activity, often confusing its purpose and blurring social intentions.

Man's religious, artistic, and psychological needs in the first stages of civilization building somewhere near 3000 B.C.—at least 5000 years after the final recession of the ice—reveal his clouded understanding. The small semidemocratic urban states of Sumer were gradually absorbed by such absolutisms as Assyria. Many an Egypt throughout the ancient world displaced and overwhelmed the ancient freedoms given to us as a heritage by those independent, wandering Ice Age pioneers. This ebb and flow of civilizations occurs constantly throughout the world.

Throughout history we can trace the gunbursts of creativity and progress experienced by small populations of highly intelligent citizens, followed by an expanding population taking advantage of the momentarily accumulated surpluses. Their social and cultural decline, accompanied by power grabs, were always undergirded by mythological religious persuasions. As the Middle East stagnated, the Greek city-states stepped front and center, taking the techniques and advances of their predecessors and molding them into a unique national amalgam. In about 1500 B.C., wandering down from the northern forest, these Indo-European migrants added an exhilarating element of intelligence and spontaneity to the declining Near Eastern civilizational scene.

The perfection of the phonetic alphabet, the self conscious enunciation of democratic political and social life, the separation of religion from secular thought, and the creation of philosophy were only a small part of the contribution to mankind by this astounding ethnic group. Science too was created by the Greeks, though never systematized. Perhaps Roman centralization and industrial slavery undermined the progress toward a modern science made by the later Hellenistic Greeks.

What we do see in this culture is the ability of the Greeks to distill man's various symbolic mental interests and to pour this awareness into creative activity in each domain. Witness the sculpture, architecture, a supreme lithic industry, often produced by largely unknown but talented artists. Drama, both tragedy and comedy, were raised to a peak of creativity.

Let us look back now at the Greeks carving from their rocky land a small, unique civilization. In studying them, we can better understand the mission of human nature, as illustrated by the blind forces of natural selection. Earn enough bread so that the mind is free to play the game of "excess neurons," all that restless mental agitation that culminates in the rich web of civilizational meaning. The Greek philosopher Aristotle intuited this need of sapient man. Intelligent humans, he said, work in order to satisfy body needs, to provide time and leisure for the mind to pursue its interests—and what diverse and intellectually deep interests they were! The fruits of human potentiality are there in Greece for all to see.

The Romans built on Greek achievements, but for various reasons they took a different tack. They explored power and expansion. A worldwide civilization was put together practically, opportunistically, with a superb extension of the military machine that was first demonstrated by the Greeks. This practicality had, however effaced the theoretical and creative genius of the Greeks. The

Romans were mired in the world of immediate goals and dreams. They did not envision the ultimate exhaustion entailed by power, wealth, and the sober mind.

The Greek philosophical vision had separated the religious element from the sense of the unknown, the possible future, that is contained in speculative thought. No gods here to limit or direct speculation about man's place in nature. Philosophy was a thought platform leading us toward an unseen reality.

Unfortunately the Greeks were not able to complete the scientific revolution that first would have disciplined this imaginative thought with deductive logic and then empirical testing. Roman realism here effaced Greek eros. They saw the needs, the future, the possible, straight ahead. Their sculptured busts, mostly crafted by Greek artists, reflect their sober, down-to-earth vision.

Then Christianity. One of many mystery religions contending for man's nonpractical emotional social needs, it absorbed the more powerful political and intellectual elements of this international world of peoples and cultures. From the Greeks, philosophy, from the Jews monotheistic messianism, from the Romans practical, organizational, diocesan politics. As Rome waned, Christianity, as in the process or petrification, seeped into the pores, a great otherworldly international movement.

Now a pause, another thousand years, during which northern Europe endured hardship, darkness, and travail. In the fallow soil a new civilization had been incubating. The fog of mysticism had been solidified into an institution that had learned to cope with a practical world throbbing with creative energy. The Church, now porous and itself aging and infirm, allowed the ferment to take hold.

Two factors seemed to release anew human intelligence: First, an environment of heterogeneous small city-states and nations existed whose power was minimal and in which the individual was relatively free. Second, the knowledge of the Greeks and Romans was now largely recovered, the relationship between knowing and doing, thinking and working was secured by the necessities of survival. Science was the product, and along with science a burst of creative endeavor in every area of cultural discourse that rivaled the Hellenic tradition of two thousand years earlier.

It was not long before science begot technology and eventually the realization that a method of thought had been produced that could systematically uncover the unseen structure of physical and biological experience, possibly even of our social world. The enlightened and optimistic vistas produced from the seventeenth and eighteenth centuries were probably unique to human experience in that they were international; they were shared not merely by one people speaking one language—as with the Greeks—but were part of an international experience that could be shared by all mankind.

The consequences are apparent. Human intelligence operating out of western Europe and North America led the way out of the scientific revolution into the industrial revolution of the nineteenth century and beyond. The culture of intelligence that burst from all sectors of the compass rejected the skepticism of ten thousand years of cultural ebb and flow, the chastening of human history. In the wake came a river of peoples, moving over the face of this earth out of their homelands and proliferating in numbers beyond the scope of our experience and eventually beyond the carrying power of the globe.

Human intelligence, first highly selected when man used it to compete with his progenitor apes and monkeys, then to war against his cousins and brothers, had always had its adaptive functions cut out for itself. To preserve itself, mankind had to look beyond the horizon, to integrate the clues from yesterday's and today's experience into an equation that would anticipate tomorrow. This was the message of the cortex: find the means to understand so as to anticipate. Written language was the first stage of objectification, philosophy the next. Science was the fulfillment of this possibility.

It is a truism now in the late twentieth century that our knowledge and mastery of the physical, even the biological, worlds have outrun our understanding of man himself. The promising social changes that began in the nineteenth century led to a series of genocidal episodes in the twentieth that destroyed much of the genetic potential for intellectual advance. At the same time, advances in agriculture, disease control, even technological competency stimulated an enormous eruption of peoples from all parts of the globe, especially from the southern latitudes, peoples who were less ready to institutionalize the economic and social disciplines so necessary for cultural advance.

The truth is, as rapidly as humankind was becoming a truly panmictic species (crossbreeding of the various historical races), the upsurge of population was probably diluting the intellectual levels that had made possible this perilous advance in the first place. Anthropologists speak of a decline in average endocranial capacity of up to 200 cm³ as compared to the late-Pleistocene Cro-Magnon peoples. The traditional bell-shaped curve of intelligence testing reveals the unevenness in academic potential for learning those high cognitive skills so essential to the demands of modern life. The general debasement of contemporary culture, as compared even with that of the eighteenth or nineteenth century, echoes the consequence of this undisciplined population explosion.

Furthermore, science and technology have not produced the larger philosophical plan that would guide our species to master events and lead a course of human improvement into the next century. The irony is that just as we are on the verge of eliminating brute labor in order to achieve high social standards of life, of controlling disease, pestilence, and want, the demographic and cultural dynamics have come crashing down on mankind's illusions of freedom and hope.

Given the uneven educational and cultural levels worldwide, it is not surprising that the snake oil salesmen of ideology and myth have sent up their perfumed smokescreens to confuse the masses. What to think, how to act, and where to go seem to be hopeless questions.

Where will intelligence lead our species? What ought we think about the message of this evolutionary process? Certainly it must encompass more than the clever tricks of space exploration or moon walking. Certainly it will necessitate the discipline and clear thinking that have distilled the varied levels of human thought and symbolism in civilized living. Religion and myth, the aggressive enthusiasms of linguistic and ethnic tribalism that we carry with us from the primeval past, all have a place in the human pantheon. For purposes of clarity, however, they need to be separated from the social vision and planning; the philosophical programs must be effected without the dogmas of limbic system importunings.

The models of human societies are there. While the past cannot be a rigid determiner of future right action, we have from those examples—Sumer, Hellenic Greece, Renaissance and early modern Europe, and North America—insights into the creative lives of free peoples in small self-governing communities. Here intelligence was sober, instrumental, yet it released the powerful creative energies that make life worth living.

In northeast Asia—China, Korea, and Japan—a reservoir of intelligence now exists that probably stands above the wounded west. These peoples have enormous demographic problems, also tendencies to enmesh themselves in hallucinatory ideologies and visions of power. But the intelligence is there; the historical moment seems to beckon them forward.

The Mongoloids were late in receiving that infusion of highly sapient genes that precipitated their final upward selective surge. They were late in receiving and using the culture and literacy that also came from the west. Though they used them well to develop their own particular civilization, they held back. They did not join the west.

Having a less mythological religious tradition, the Confucian ethical plan could have led to a step-by-step advance and eventual parity with the west, but the almost medieval bureaucracies were unwilling to join the fray. Eventually, in the late nineteenth century, events came to them; they could no longer turn away. A century later, having experienced the tumult and psychosis of modernization, they seem to be moving. Should we not look in the direction of this new leadership segment for a rational message for mankind? Should we not be looking for guidance from a population of high intelligence, education, and ability, and one that is now willing?

In the more distant future, the images are clouded. The chaos of the developing world, the west's loss of nerve, the shadowy grip of totalitarianisms that seem to be spreading irrevocably have only one unarguable antidote. What resolved the chaos of decline in Rome was deprivation. The European world went back to square one. If internal social dissolution does not sweep the historical board clean, new ecological conditions may.

The triumph of the intelligent has not been unilinear. Hardship has always been prelude to a new lease, a quantum jump in structure and possibilities. These past forty-five to fifty thousand years during which *Homo sapiens sapiens* made his mark on the world have inflicted scarring wounds—on the ecology, on the flora and fauna—the side effects of "progress." Now, the price of renewal may be unbearably high.

# Bibliography

Chapter I: **About the Unspeakable.**

A sample of books in the tradition of man
and his animal adaptations.

Ardrey, Robert. 1976. *The Hunting Hypothesis*. New York: Atheneum.

Lorenz, Konrad. 1963. *On Aggression*. New York: Harcourt, Brace and World.

Lumsden, Charles and Wilson, Edward O. 1983. *Promethean Fire*. Cambridge: Harvard University Press.

Reynolds, Peter. 1981. *On the Evolution of Human Behaviour: the argument from animals to man*. Berkeley: University of California Press.

Tiger, Lionel. 1970. *Men in Groups*. New York: Random House.

Wilson, Edward O. 1979. *On Human Nature*. Cambridge: Harvard University Press.

Chapter II: **Intelligence in Search of An Animal.**

Recent Classics in the Origin of Life.

Blum, Harold. F. 1951. *Time's Arrow and Evolution*. Princeton, New Jersey: Princeton University Press.

Cloud, P. 1982. "How Life Began." *Nature*, March, Vol. 296, No. 5854, 198-199.

Crick, Francis. 1982. "Life Itself: Its Origin." *Nature*, April, Vol. 296, No. 5857, 496-497.

Luria, S. E. 1973. *Life the Unfinished Experiment.* New York:    Charles Scribner's Sons.

Maddox, John. 1983. "Simulating the Replication of Life." *Nature.* October 6, Vol. 305:    469.

Nigrelli, Ross F. 1957. "Modern Ideas on Spontaneous Generation." New York:    *New York Academy of Sciences,* August 30, Vol. 69, Art. 2.

Oparin, A. I. 1953. *Origin of Life.* New York:    Dover (1938).

Oparin, A. I., ed. 1959. *The Origins of Life on Earth.* New York:    Pergamon Press.

Wolman, Y. 1981. *Origin of Life.* Boston:    E. Reidel.

Intelligence and the Evolutionary Process.

Eldridge, Niles and Gould, Stephen J. 1972. "Punctuated Equilibrium: An Alternative to Phyletic Gradualism" in Schopf, T. J. M., ed. *Models in Paleobiology.* San Francisco:    Freeman, Cooper, 82-115.

Halstead, W. C. 1947. *Brain and Intelligence.* Chicago:    University of Chicago Press.

Jerison, Harry. 1973. *Evolution of the Brain and Intelligence.* New York:    Academic Press.

Macphail, E. M. 1982. *Brain and Intelligence in Vertebrates.* New York:    Oxford University Press.

Mayr, E. 1963. *Animal Species and Evolution.* Cambridge:    Harvard University Press.

Simpson, G. G. 1949. *The Meaning of Evolution.* New Haven:    Yale University Press.

Stenhouse, David. 1973. *The Evolution of Intelligence.* New York:    Harper and Row.

Chapter III: **Odd Anthropoid.**

Bakker, Robert. 1978. "Dinosaur Renaissance," in *Evolution and the Fossil Record. Scientific American*, 125-141. San Francisco: W. H. Freeman.

Campbell, Bernard, G. 1960. *Human Evolution.* Chicago: Aldine Publishers.

Ciochin, Russell and Corracini, R. S., eds. 1983. *New Interpretations of Ape and Human Ancestors.* New York: Plenum.

Colbert, E. H. 1973. *Wandering Lands and Animals.* New York: E. P. Dutton.

Goodman, M. et al. 1983. "Evidence on Human Origins from Haemoglobin of African Apes." *Nature.* June 9 Vol. 303, 546-548.

Gribbin, John and Cherfas, Jeremy. 1982. *The Monkey Puzzle.* New York: Pantheon.

Kemp, T. S. 1982. *Mammal-Like Reptiles and the Origin of Mammals.* New York: Academic Press.

LaBarre, Weston, Hockett, C. F. and Ascher, R. 1964. "The Human Revolution." *Current Anthropology.* Vol. 3, 135-168.

Lewin, Roger. 1983. "Is the Orangutan a Living Fossil?" *Science.* December 16, Vol. 222.

Pilbeam, David. 1982. "New Hominoid Skull Material from the Miocene of Pakistan." *Nature.* January 21, Vol. 295, 232-233.

Sarich, Vincent. 1968. "The Origin of the Hominids: An Immunological Approach," in Washburn, S. L. and - Jay, P. C., eds. *Perspectives on Human Evolution.* Vol. 94.

Smith, Homer. 1961. *From Fish to Philosopher.* Garden City, New York: Doubleday.

Yunis, J. J. and Prakash, O. 1982. "The Origin of Man: A Chromosomal Pictorial Legacy." *Science.* March 19, Vol. 215.

Chapter IV: **First Crisis and the Birth of** *Homo.*

Clark, J. D. 1984. "Paleoanthropological Discoveries in Middle Awash Valley, Ethiopia." *Nature.* February 2, Vol. 307, 423-428.

Ciochin, Russell and Corracini, R. S., eds. 1983. *New Interpretations of Ape and Human Ancestors.* New York: Plenum.

de Beer, Sir Gavin. 1958. *Embryos and Ancestors.* Oxford: Oxford University Press.

Gould, Stephen Jay. 1977. *Ontogeny and Phylogeny.* Cambridge: Harvard University Press.

Harding, R. S. O. and Teleki, G., eds. 1981. *Omnivorous Primates.* New York: Columbia University Press.

Lewin, Roger. 1983. "Is the Orangutan a Living Fossil?" *Science.* December. Vol. 222.

Lovejoy, C. Owen. 1981. "The Origin of Man." *Science.* January 22, Vol. 211, 341-350.

Passingham, Richard. 1982. *The Human Primate.* San Francisco: W. H. Freeman.

Pilbeam, David. 1982. "New Hominoid Skull Material from the Miocene of Pakistan." *Nature.* January 21, Vol. 295, 232-233.

Raff, R. A. and Kaufman, T. C. 1983. *Embryos, Genes and Evolution: The developmental-genetic basis of evolutionary change.* New York: Macmillan.

Szalay, Frederick S. and Delson, Eric. 1979. *Evolutionary History of the Primates.* New York: Academic Press.

Chapter V: **Breakout.**

Andrews, Peter and Cronin, J. F. 1982. "The Relationship of *Sivapithecus* and *Ramapithecus* and the Evolution of the Orangutan." *Nature.* June, Vol. 297, 541-546.

Boaz, N. T., Howell, F. C., and McCrossin M. C. 1982. "Faunal Age of the Usno, Shungura B and Hadar Formations, Ethiopia." *Nature.* December 16, Vol. 300.

Bradshaw, John L. and Nettleton, Norman. 1983. *Human Cerebral Asymmetry.* New York: Prentice Hall.

Brain, C. K. 1981. *The Hunter and the Hunted.* Chicago: University of Chicago Press.

Day, M. H. 1982. *"Lucy* Jilted." *Nature.* December 16, Vol. 300, 574.

Holloway, Ralph. L. 1974. "The Casts of Fossil Hominid Brains." *Scientific American.* July, 106-115.

Holloway, Ralph. L. 1976. "A New Study: Brain Mold in Old Skull: Skull 1470." *New York Times.* April 21, 33, 62.

Holloway, Ralph. L. et al. 1982 "Endocast Asymmetry in Pongids and Hominids." *American Journal of Physical Anthropology.* Vol. 58, 101-110.

Hooton, Ernest. 1946. Rev. Ed. *Up from the Ape.* New York: Macmillan.

Johanson, Donald C. 1981. *Lucy.* New York: Simon and Schuster.

Leakey, Richard. 1973. "Skull 1470—New Clue to Earliest Man?" *National Geographic.* June, Vol. 143, No. 6, 819-829.

Lewin, Roger. 1983. "Were Lucy's Feet Made for Walking?" *Science,* May 13, Vol. 220, 700-702.

Pilbeam, David. 1972. *The Ascent of Man.* New York: Macmillan.

Pilbeam, David and Jacobs, L. L. 1978. "Changing Views of Human Origins." *Plateau Quarterly.* Vol. 51, No. 1, 18-31.

Chapter VI: **The Painful Truth.**

Davis, Dan. D. 1981. *The Unique Animal: the Origin, Nature and Consequences of Human Intelligence.* New York: Prytaneum Press.

Diamond, Jared. 1982. "Man the Exterminator." *Nature.* August 26, Vol. 298, 787-789.

Howell, F. Clark. 1965. *Early Man.* New York: Time-Life, Life Nature Library.

Leakey, Richard and Lewin, Roger. 1977. *Origins.* New York: E. P. Dutton.

von Koenigswald, G. H. R. 1962. *The Evolution of Man.* Ann Arbor: University of Michigan Press.


Chapter VII: **Homo Stabilis: The Model of Man.**

Binford, Lewis R. 1981. *Ancient Men and Modern Myths.* New York: Academic Press.

Coon, Carleton. 1966. "The Taxonomy of Human Variation." *Annals of the New York Academy of Sciences.* February 28, Vol. 134, Article 2, 516-523.

Eiseley, Loren. 1957. *The Immense Journey.* New York: Random House.

Fisher, Helen. 1982. *The Sex Contract.* New York: Morrow.

Harris, J. M., ed. 1983. *Koobi Fora: Researches into Geology, Paleontology and Human Origins.* Vol. 2. New York: Oxford University Press.

Oakley, Kenneth P. 1957. *Man the Tool Maker.* Chicago: University of Chicago Press.

Orquera, Luis Abel. 1984. "Specialization and the Middle/Upper Paleolithic Transition." *Current Anthropology*. February, Vol. 25, No. 1, 73-98.

Shapiro, Harry L. 1974. *Peking Man*. New York:   Simon and Schuster.

Tattersall, Ian and Delson, Eric. 1984. *Ancestors*. New York:   American Museum of Natural History.


Chapter VIII: **The Brain Just Grew**.


Dobzhansky, Theodosius. 1962. *Mankind Evolving*. New Haven:   Yale University Press.

Eccles, John C. 1973. *The Understanding of the Brain*. New York:   McGraw Hill.

Rensch, Bernhard. 1959. *Evolution Above the Species Level*. New York:   Columbia University Press.

Russell, E. S. 1946. *The Directiveness of Organic Activities*. London:   Cambridge University Press.

Simpson, George G. 1953. *The Major Features of Evolution*. New York:   Columbia University Press.


Chapter IX: **The Children of Natural Selection**


Binford, Lewis R. 1983. *In Pursuit of the Past: Decoding the Archeological Record*. New York:   Thames and Hudson.

Coon, Carleton and Hunt, Edward E., Jr. 1965. *The Living Races of Man*. New York:   Alfred Knopf.

Dawkins, Richard. 1976. *The Selfish Gene*. New York:   Oxford University Press.

Dawkins, Richard. 1982. *The Extended Phenotype: the Gene as a Unit of Selection*. San Francisco:   W. H. Freeman.

Eysenck, Hans J. and Eysenck, Michael. 1981. *Mindwatching*. Garden City, New York: Anchor, Doubleday.

Leakey, Richard and Lewin, Roger. 1977. *Origins*. New York: E. P. Dutton.

Moorehead, Alan. 1966. *The Fatal Impact*. New York: Harper and Row.

## Chapter X: **The Erectine Finale.**

Colbert, Edwin H. 1975. "Mammoths and Men," in *Ants, Indians, and Little Dinosaurs*, ed. Alan Ternes. New York: Charles Scribner's Sons, pp. 180-188.

Day, M. H. 1973. *Human Evolution*. New York: Barnes and Noble.

Howell, F. C. 1965. *Early Man*. New York: Time-Life, Life Nature Library.

Martin, Paul S. 1975. "Pleistocene Overkill," in *Ants, Indians, and Little Dinosaurs*, ed. Alan Ternes. New York: Charles Scribner's Sons, pp. 189-199.

Pilbeam, David. 1972. *The Ascent of Man*. New York: Macmillan.

Poulianos, A. N. 1977. "Protoeuropeoid Man and His Origins." *Journal of Human Evolution.*" Vol. 6, 259-261.

Shapiro, Harry. 1974. *Peking Man*. New York: Simon and Schuster.

Tobias, Phillip V. 1980. *"Homo Habilis* and *Homo Erectus."* *Anthropologie* (Brno). Vol. 18, 2-3.

von Koenigswald, G. H. R. 1962. *The Evolution of Man*. Ann Arbor: University of Michigan Press.

Washburn, S. L. and Dolhinow, Phyllis, eds. 1972. *Perspectives on Human Evolution*. New York: Holt, Rinehart and Winston.


Chapter XI: **Into the Tunnel.**


Bolk, L. 1926. *Das Problem der Menschenwerdung*. Jena.

de Beer, Sir Gavin. 1968. *Embryos and Ancestors*. New York: Oxford University Press.

Eldridge, Niles and Gould, S. J. 1972. "Punctuated Equilibrium: An Alternative to Phyletic Gradualism." in Schopf, T. J. M., ed. *Models in Paleobiology*. San Francisco: Freeman, Cooper. 82-115.

Gould, S. J. 1977. *Ontogeny and Phylogeny*. Cambridge: Harvard University Press.

Gould, S. J. 1982. "Darwinism and the Expansionism of Evolutionary Theory." *Science*, April 23, Vol. 216, 380.

Gould, S. J. 1984. "Balzan Prize to Ernst Mayr." *Science*, January 20, Vol. 223, 255-257.

Gregory, W. K. 1937. "Supra-specific Variation in Nature and in Classification; a Few Examples from Mammalian Paleontology." *American Naturalist*, Vol. 71, 268-276.

Griffiths. 1978. *The Biology of the Monotremes*.

Itzkoff, Seymour. W. 1983. *The Form of Man, the evolutionary origins of human intelligence*. Ashfield, Massachusetts: Paideia Publishers.

Mayr, Ernst et al. 1982. "Punctuationism and Darwinism Reconciled?" *Nature*, Vol. 296, 608.

Rhodes, E. H. T. 1983. "Gradualism, Punctuated Equilibrium and the Origin of the Species."*Nature*, April 23, Vol. 216.

Raff, R. A. and Kaufman, T. C. 1983. *Embryos, Genes and Evolution: the developmental-genetic basis of evolutionary change.* New York: Macmillan.

Wright, Sewall. 1963. "Adaptation and Selection," in Jepson, G. L., Simpson, G. G., Mayr, Ernst, eds. *Genetics, Paleontology and Evolution.* New York: Atheneum, 365-391.

## Chapter XII: Tipping the Scales.

Bordes, F., ed. 1972. *The Origin of Homo Sapiens.* Paris: UNESCO.

Conroy, Glenn et al. 1978. "Newly Discovered Fossil Hominid Skull from the Afar Depression, Ethiopia." *Nature,* November, Vol. 276, No. 2, 67-70.

Coon, Carleton. 1962. *The Origin of Races.* New York: Alfred Knopf.

Day, M. H. 1970. *Fossil Man.* New York: Grosset and Dunlap (Bantam Book).

Day, M. H. 1977. *Guide to Fossil Man.* Chicago: University of Chicago Press.

Day, M. H. and Leakey, M. D. 1980. "A New Hominid Fossil Skull (L. H. 18) From the Nguloba Beds, Laetoli, Northern Tanzania." *Nature,* March 6, Vol. 284, 55-56.

Goldstein, M. 1982. "The Lower to Middle Paleolithic Transition and the Origin of Modern Man." *Current Anthropology.* February, Vol. 23, No. 1, 124-126.

Jacob, Teuku. 1979. "Hominine Evolution in South East Asia." *Archaeology and Physical Anthropology in Oceania,* April, Vol. 14, No. 1, 1-10.

Kennedy, Gail. 1980. "The Emergence of Modern Man." *Nature,* March 6, Vol. 284, No. 11.

Pilbeam, David. 1972. *The Ascent of Man*. New York: Macmillan.

Rightmire, Philip. 1983. "The Lake Ndutu Cranium and early *Homo Sapiens* in Africa." *American Journal of Physical Anthropology*, Vol. 61, 245-254.

## Chapter XIII: The Neanderthal Experiment.

Binford, S. R. 1968. "Early Upper Pleistocene Adaptation in the Levant." *American Anthropologist*, Vol. 70, 707.

Brose, D. S. and Wolpoff, M. H. 1971. "Early Upper Paleolithic Man. . ."*American Anthropologist*. Vol. 73, 1156-94.

Day, M. H. 1973. *Human Evolution*. New York: Barnes and Noble.

Dinnell, Robin. 1983. "A New Chronology for the Mousterian." *Nature*, January 20, Vol. 301, 199.

Howells, W. W. 1974. "Neanderthals: Names, Hypotheses, Scientific Method." *American Anthropologist*. Vol. 76, 24-38.

Jelinek, Jan. 1971. "On Early Man in Central and Eastern Europe." *Current Anthropology*, Vol. 12, 2: 241-242.

Jelinek, Jan. 1982. "The Tabun Cave and Paleolithic Man in the Levant." *Science*, January 25, Vol. 216, 1369.

Kurtén, Björn. 1972. *Not from the Apes: The History of Man's Origins and Evolution*. New York: Vintage Books.

Lieberman, Philip. 1976. "More Talk on Neanderthal Speech." *Current Anthropology*, June, Vol. 19, 2: 407.

Solecki, Ralph. S. 1960. "Three Adult Neanderthal Skeletons from Shanidar Cave in Northern Iraq." *Smithsonian Report Publications for 1959*. No. 4414, 603-635.

Solecki, Ralph. S. 1975. "Shanidar IV, a Neanderthal Flower Burial in Northern Iraq." *Science*, 190, 880-881.

Trinkaus, Erik. 1983. *The Shanidar Neanderthals.* New York: Academic Press.


## Chapter XIV: **Man of the Future.**

Baker, John. 1974. *Race.* New York: Oxford University Press.

Coon, Carleton. 1962. *The Origin of Races.* New York: Alfred Knopf.

Coon, Carleton. 1965. *The Living Races of Man.* New York: Alfred Knopf.

Eibl-Eibesfeldt, Iraneus. 1974. "The Myth of the Aggression-Free Hunter and Gatherer Society," in Holloway, Ralph, ed. *Primate Aggression, Territoriality and Xenophobia.* New York: Academic Press.

Eiseley, Loren. 1957. *The Immense Journey.* New York: Random House.

Schenk, Gustave. 1961. *The History of Man.* New York: Chilton Company—Book Division.

Thomas, Elizabeth. 1959. *The Harmless People.* New York: Alfred Knopf.

Tobias, Phillip. 1971. *The Brain in Hominid Evolution.* New York: Columbia University Press.


## Chapter XV: **The Middle Kingdom Moves Outward.**

Brues, Alice. 1977. *People and Races.* New York: Macmillan.

Darlington, C. D. 1969. *The Evolution of Man and Society.* London: Allen and Unwin.

Itzkoff, Seymour W. 1983. *The Form of Man, the evolutionary origins of human intelligence.* Ashfield, Massachusetts: Paideia Publishers.

Phenice, T. W. 1972. *Hominid Fossils*. Dubuque, Iowa: Wm. C. Brown.

Shapiro, Harry. 1974. *Peking Man*. New York: Simon and Schuster.

Weidenreich, Franz. 1946. *Apes, Giants and Man*. Chicago: University of Chicago Press.

Yi, Sionbok and Clark, G. A. 1983. "Observations on the Lower Paleolithic of Northeast Asia." *Current Anthropology*, April, Vol. 24, 181-202.


### Chapter XVI: **Ultimate Sapiens: Cro-Magnon.**

Bahn, Paul G. 1983. "A Paleolithic Treasure House in the Pyrenees." *Nature*, April 14, Vol. 302.

Bailey, G., ed. 1983. *Hunter-Gatherer Economy in Pre-history: A European Perspective*. New York: Cambridge Press.

Budko, V. D. 1972. "The Paleolithic Period of Byelorussia," in *The Origin of Homo Sapiens*, ed. by F. Bordes. 187-198. Paris: UNESCO.

Clark, Grahame. 1967. *The Stone Age Hunters*. New York: McGraw Hill.

Clark, Grahame and Piggott, Stuart. 1967. *Pre-Historic Societies*. New York: Alfred Knopf.

Gamble, Clive. 1980. "Information Exchange in the Paleolithic." *Nature*, February 7, Vol. 283, 522.

Leakey, Richard. 1981. "The Making of Mankind." *The Listener*, Spring, parts 1-7.

Oakley, Kenneth P. 1957. *Man the Tool Maker*. Chicago: University of Chicago Press.

Pfeiffer, J. E. 1983. *The Creative Explosion: An Inquiry Into the Origins of Art and Religion*. New York: Harper and Row.

Schild, Romuald. 1976. "The Final Paleolithic Settlements of the European Plain." *Scientific American*, February, Vol. 234, 2: 88-99.

Zvelebil, Mark. 1984. "Clues to Recent Human Evolution from Specialized Technologies." *Nature*, January 26, Vol. 307, 314-315.


## Chapter XVII: **Thinking Man**.

Cassirer, Ernst. 1953-1957. *The Philosophy of Symbolic Forms.* 3 Volumes. New Haven: Yale University Press. (1923-1929).

Itzkoff, Seymour W. 1971. *Ernst Cassirer: Scientific Knowledge and the Concept of Man.* Notre Dame: University of Notre Dame Press.

Lévi-Strauss, Claude. 1966. *The Savage Mind.* Chicago: University of Chicago Press.

Littauer, M. A. and McCarthy, F. D. 1974. "On Upper Paleolithic Engraving." *Current Anthropology*, September, Vol. 15, 3: 327-328.

Marshack, Alexander. 1972. *The Roots of Civilization.* New York: McGraw-Hill.

Marshack, Alexander. 1972. "Cognitive Aspects of Upper Paleolithic Engraving." *Current Anthropology*, June-October, Vol. 13, 3-4: 445-477.

Marshack, Alexander. 1979. "Upper Paleolithic Symbol Systems of the Russian Plain: Cognitive and Comparative Analysis." *Current Anthropology*, June, Vol. 20, 2: 271-311.

Chapter XVIII: **Intelligence at War with Human Nature.**

Note: The parable that begins this chapter is evoked by the memory of the Sumerian cities in which literacy and the arts of civilization began to flourish as early as 4000 B.C. There are subsequent suggestions of the existence of a limited democratic monarchical political structure, perhaps reminiscent of the ancient Upper Paleolithic, northern tribal patterns. At about 2400 B.C. Semitic Akkadians swept out of the desert and brought this first urban democratic civilization to an end.

Cassirer, Ernst. 1946. *The Myth of the State.* New Haven: Yale University Press.

Koestler, Arthur. 1967. *The Ghost in the Machine.* New York: Macmillan.

MacLean, P. D. 1973. *A Triune Concept of the Brain and Behaviour.* Toronto: University of Toronto Press.

Sagan, Carl. 1977. *The Dragons of Eden.* New York: Random House.

Simeons, A. T. W. 1960. *Man's Presumptuous Brain.* New York: Dutton.

# Index

## A

Abbevillian, hand ax, 74

Abstraction, 159-160, 166-167; symbolization and, 160-162, 179

Acheulean culture, 72, 74-75, 99, 123, 125, 144, 153

Adaptation: of *Homo*, 19, 90; intelligence as an, 18; life forms, 25-35; and plains life, 68

*Aegyptopithecus*, 49; *see also* Apes

Africa: origin of *Homo*, 61, 66, 113ff.; erectine homeland, 75, 101, 143-144; and Neanderthal sapiens, 123

Aggression in *Homo*, 20, 63ff., 80, 93

Ainu, 142; *see also* Hybridization in man

Altamira, 113, 154

American Museum of Natural History, 72, 73

Amino acids, 57

Amniotic egg, 34, 42

Amphibians, 42

Andaman Islanders, 142; *see also* Race: Australids

Animals: evolutionary origins, 25-27; adaptation, 27, 33, 97

Apes: call system, 38; chimpanzee, baboon, 103; decline in numbers, 20, 40, 65-67, Diagram 4; and human development, 108; orangutan, 63-64; origins, 36-38, 49; onto the plains, 55-59

Aphrodite, *see* Sexuality

Aristotle, 180

Art: Boskop, Bushman, 134-135, 138-139, Plate 3; origins of Chinese, 148; Cro-Magnon peoples, 151-158, Plate 1, 2

Asia, Northeast, *see* Race: Mongoloids

Assyria, 179

Aterians, 135-136; *see also* Race: Capoids, Boskopoids

Aurignacian culture, 153; *see also Homo sapiens sapiens*: Cro-Magnon

Australopithecines, 20, 22, 49, 56, 59-62, 65-66, 70-71

## B

Behaviorism, 80

Biochemistry of anthropoids, 37-38

Bipedalism, 84, 89, 117

Bodo, 116; *see also Homo erectus*: African borderline

Boskopoids, 133-140

Brain: ape and human, 38-39; and civilization, 163-165; dietary needs of human, 68; direction of evolution of, 81-85, 87-88, 99; hemispheric specialization in,